Curious Obsessions

Rachael Kohn

Curious Obsessions

In the History of Science and Spirituality

Rachael Kohn

.

Adelaide
2020

A Forum for Theology in the World
Volume 7, Issue 1, 2020

A Forum for Theology in the World is an academic refereed journal aimed at engaging with issues in the contemporary world, a world which is pluralist and ecumenical in nature. The journal reflects this pluralism and ecumenism. Each edition is theme specific and has its own editor responsible for the production. The journal aims to elicit and encourage dialogue on topics and issues in contemporary society and within a variety of religious traditions. The Editor in Chief welcomes submissions of manuscripts, collections of articles, for review from individuals or institutions, which may be from seminars or conferences or written specifically for the journal. An internal peer review is expected before submitting the manuscript. It is the expectation of the publisher that, once a manuscript has been accepted for publication, it will be submitted according to the house style to be found at the back of this volume. All submissions to the Editor in Chief are to be sent to: hdregan@atf. org.au.

Each edition is available as a journal subscription, or as a book in print, pdf or epub, through the ATF Press web site — www.atfpress.com. Journal subscriptions are also available through EBSCO and other library suppliers.

Editor in Chief
Hilary Regan, ATF Press

A Forum for Theology in the World is published by ATF Theology and imprint of
ATF (Australia) Ltd (ABN 90 116 359 963) and
is published twice or three times a year.

ISBN: 978-1-925679-84-7 soft
       978-1-925679-85-4 hard
       978-1-925679-86-1 epub
       978-1-925679-87-8 pdf

Published by:

An imprint of the ATF Press Publishing
Group owned by ATF (Australia) Ltd.
PO Box 234
Brompton, SA 5007
Australia
ABN 90 116 359 963
www.atfpress.com
Making a lasting impact

About the Author

Rachael Kohn is a Religious Studies academic and broadcaster whose first book, *The New Believers: Reimagining God* (2004) together with her 'contributions to society' earned her a Doctor of Letters *honoris causa* in 2005 from the University of New South Wales.

From 1992–2018 Rachael produced and presented religion programs for ABC Radio National, including Religion Report, Religion Today, The Ark, and the long running Spirit of Things, as well as documentaries for radio and TV for which she won international awards including 2 World Gold Medals.

In 2019 Rachael Kohn was awarded the Order of Australia and in the same year she was made Fellow of the Royal Society of New South Wales.

Table of Contents

Acknowledgments

I wish to thank the many brilliant historians who have provided the fascinating stories from our religious past which were the basis of The Ark, a program I produced and presented for ABC Radio National for six years. They inspired this book but of course do not bear any responsibility for it; they enlarged my own knowledge gleaned from a lifetime researching and teaching religious studies. The thematic synthesis into the essays comprising *Curious Obsessions* is wholly my own.

I would like to thank Ali Lavau and Susan Morris-Yates for their expert oversight of the original publication for ABC Books (later Harper Collins), and Hilary Regan and Alan Cadwallader of ATF Press for encouraging me to reprise the contents of this book for the international theological community.

Preface to the Revised Edition

'We live in unprecedented times,' is a commonly heard phrase whenever we are faced with difficult challenges. Yet history almost always proves us wrong. Even the current worldwide pandemic that has so far affected more than a million people, is one in a long line of communicable diseases that have wreaked havoc on large populations. Some things have changed of course. By comparison to the Spanish flu which killed between 20 and 50 million people and the Bubonic plague which five centuries earlier killed approximately 25 million, the progress of medical science and hospital care has ensured that a far smaller percentage of people who contract the COVID-19 coronavirus will actually die from the disease.

Yet despite the advances in science and the knowledge we have of where and how the disease was started in China, there is an alarming revival of age-old conspiracy theories about the source and spread of the disease. 'Accusations that Jews created coronavirus, that affected Jews are being punished, and calling COVID-19 the "Jew flu" are just some examples of anti-Semitism expressed online by Australians during the current pandemic.'[1] At present a raft of conspiracy beliefs circulating on the internet blames the Jews for the deadly virus and the subsequent economic hardship it has caused. Muslim nations, such as Iran, Turkey and Pakistan[2] claim that Israel is behind the worldwide spread of COVID-19. This classic anti-Semitic trope,

---

1. https://ajn.timesofisrael.com/COVID-19-anti-Semitism-reaches-australia/
2. https://www.haaretz.com/world-news/premium-against-coronavirus-pakistan-turns-to-a-traditional-remedy-blame-ahmadis-and-jews-1.8902949
   https://cst.org.uk/data/file/d/9/Coronavirus%20and%20the%20plague%20of%20antisemitism.1586276450.pdf

which also attributed the Bubonic plague, the AIDS virus and the Ebola epidemic to the Jews, and is promoted by a well-known author, David Icke (see chapter 7), whose books are sold on Amazon, is one of the many reasons that the study of history is so important in our age.

Disease, race, medicine and religion, are a heady mix but combined with disaster predictions and utopian dreams they are among the curious obsessions in the history of science and spirituality which have animated our past and found new life in the present. Today anti-vaxxers, such as the infamous Dr Mikovitz of the *Plandemic* video,[3] are on the rise just as they were in the early years of the twentieth century, when prominent author and playwright George Bernard Shaw was convinced that malevolent forces were behind the novel method of preventing disease (see chapter 5). Big money and shadowy cabals pulling the strings of worldwide trade are among the usual suspects in these dark imaginings.

The coronavirus crisis is just the latest in a long line of issues like those covered in this book that have attracted a wide array of scientists, prognosticators and futurists, new age spiritualists and religious thinkers. Indeed, the fascinating interplay between the spiritual urge and the scientific quest has not only had dubious outcomes but very valuable ones as well.

For four decades I have been deeply interested in the role that religion has played as a creative force in human civilisation. There is not a corner of our human endeavour that has not been infused with the spiritual quest, and that covers everything from the most intimate human relationships to man's soaring desire to glimpse the ends of the universe. From the precise layout of a sacred edifice to the inner workings of the human body, the religious mind has never had want of spaces to fill with its enthusiasms. Yet, the religious mind is often understood in quite narrow terms, as if it always indicates a traditional or pious disposition. Nothing could be less true of most of the people described in this book, all of whom were out of the ordinary, and were usually leaders rather than followers. The metaphysical dimension spurred them on to great discoveries or compelled them to realise their worldly dreams. The results were variable, from brilliant to

---

3. A 26 minute video, viewed by millions on YouTube 'claiming to present a view of COVID-19 that differs from the "official" narrative.' https://www.npr. org/2020/05/08/852451652/seen-plandemic-we-take-a-close-look-at-the-viral-conspiracy-video-s-claims

diabolical, but like any other form of human endeavour a sense of proportion is required eventually in order to succeed.

Perhaps that is why I chose to focus on curious obsessions. It is in the nature of an obsession that it lacks a sense of proportion, and because of that it exceeds the norm and breaks the boundaries, subverting the rules and the expectations of traditional authority. An astute critic might protest that, given the propensity of contemporary religion to upset tradition, there's no need to look to the past for stories of innovation. But it seems to me that it is precisely because of the tendency to view religious change as uniquely characteristic of our day that a focus on the past is necessary. At the very least, the people who are the subject of this book remind us that behind every newfangled revelation that vies for our attention today is a legion of men and women who have been there before—and their stories will astound and reward the reader with a curious mind.

Introduction

A monk who kept banned books in a secret library under the nose of the Pope; an abbess who designed extravagant royal garments for her nuns; a magician who practised the dark arts to advise Queen Elizabeth I; archaeologists who discovered the goddess just in time for feminism; utopians who never quite found what they were looking for; explorers who searched for the lost tribes of Israel but found new continents instead; a nation that was convinced it was a lost race and started a world war to prove it; an eccentric doctor and a mad monk who intuited scientific truths well before future generations would prove their theories correct . . .

There is never any shortage of astonishing stories in the history of religion. Until recently, religion was enmeshed in every area of human endeavour, including the scientific quest, social reform and the exploration of new continents. In fact, if my first book, *The New Believers, Re-imagining God*,[1] demonstrated the immensely creative approach to spirituality in the modern world, it is more than matched by the inventiveness of the geniuses and the eccentrics who peopled our past and who are the subject of this book. Would the New World have been half so interesting to Europeans had they not been convinced that the natives were Hebrew-speaking descendants of the ten lost tribes of Israel?

This belief, buttressed by linguists, historians, anthropologists, archaeologists and theologians, was not marginal in the fifteenth to the nineteenth centuries, but mainstream. Indeed, it fuelled exploration to North and South America. Later, when these lands were well

---

1. Rachael Kohn, *The New Believers: Re-imagining God* (Sydney: Harper Collins, 2003).

and truly settled by an assortment of English, Dutch, Spanish and French pioneers, it generated what is today one of the most successful religious movements in North America, the Mormons.

Similarly, the idea first put forward in the biblical book of Genesis of a quasi-angelic race of giants that walked the earth has had remarkable currency throughout the ages and has reappeared in the most unexpected places. From science fiction to New Age spirituality, the belief in a master race has gripped the imagination of founders of new religions and quite a few of their enthralled followers. When, at the end of the nineteenth century the German-speaking peoples turned their attention to their own heroic past—during one of the most unheroic periods in their history—it was to an anointed Aryan race that they looked for a destiny that would fulfil their hopes of survival and dreams of triumph. What is powerfully evident in the German example of 'the lost race' is how academic disciplines such as psychology, biology and philology conspired with religion and literature to invent a past and forge a people that would stop at nothing to claim their supreme status in Europe.

Far-fetched fantasies are easy to dismiss today as remnants of a prescientific era, until one realises that science itself is compatible with the surreal. It is in the nature of the human quest for knowledge that imagination projects as-yet-undiscovered realities (this is another way of saying it generates beliefs) in the hope of finding corroboration by convincing theories and confirmation by hard evidence. This favouring of the imagination over knowledge alone is how Einstein said he worked, and as Paul Davies has shown in his little book *The Last Three Minutes* imaginary thinking remains one of the central features of cosmological speculations.[2] Going back a few centuries, the same imaginative muse was hounding Giordano Bruno, a sixteenth-century monk who scoured the world for knowledge of the universe and everything in it. Despite the fledgling state of evidence-based science at the time, he came up with some startlingly modern notions, such as the infinity of the heavens and the atomic basis of material reality. Bruno's story as recounted here is a tragic tale of genius snuffed out, but if there is a lesson to the parable, it is that a sense of humour could have saved his life. Perhaps more difficult to achieve than is generally assumed, humour can only exist where

---

2. Paul Davies, *The Last Three Minutes*, (London: Wiedenfeld & Nicolson, 1994).

there is a capacity for ironic detachment—which, it must be said, is not always a gift of the scientific genius. It was a quality, however, which another brilliant innovator, Athanasius Kircher, cultivated to great purpose and effect, cloaking his equally devastating revelations about the nature of the universe in amusing language which even the sober Church could forgive.

Such precautions are hardly necessary today, when the scientific establishment is rarely hemmed in by the theological sensitivities of the Church. The autonomy of religion and science, into two distinct camps and removed from each other's scrutiny, has naturally prevailed in the West as a result of the separation of church and state. Conventional wisdom tells us that this is as it should be, but judging from recent trends it is by no means an absolute or secure division. Two hundred years after the Enlightenment, and on both sides of the religion and science divide, we can see sporadic movements towards the centre, specifically an urge to merge. This may indicate nostalgia for a world where religious and scientific truths were unified, a dangerously romantic view. Or it may represent a genuine deepening of our knowledge, a rediscovery of intuitive truths from the past which have trodden very close to the unsolved mysteries of the world.

Where science is concerned, I am not qualified to judge whether this move towards religion is altogether valuable, but one promising example is taken from the current annals of medicine, where researchers have raided the monastery for clues to healing the body from such varied conditions as Alzheimer's disease and rampantly growing cancer. For some impressive evidence of the former, one need only consult David Snowdon's 'Nun Study', which showed that a group of elderly nuns whose brains had extensive physiological signs of Alzheimer's suffered but few, if any, of its dysfunctions.[3] This immediately raised questions about how salutary for mental health was a life of regular prayer, tranquility and a secure community in old age. In any case, whatever medicine recommends from the world of religious practice, it is usually subject to the rigorous rules of evidence.

Material proof has a very different use, however, when the shoe is on the other foot; that is, when religious thinkers adopt scientific

---

3. David Snowdon, *Aging with Grace, What the Nun Study Teaches Us About Leading Longer, Healthier More Meaningful Lives*, (New York: Bantam, 2001); Craig Hassad, *New Frontiers of Medicine* (Melbourne: Hill of Content, 2000).

knowledge to expand their theological horizons. For them, scientific discoveries demand a theological response because the natural world, as much as the human world, is embedded with spiritual significance. Theologians who embrace the latest scientific knowledge about the cosmos do so not only out of choice, but also because they have moved on from the mythological father figure of old and are searching for other ways of understanding divinity. Contemporary theologians imagine God as Creation itself, the Life Force, and the Ground of Being, and find themselves easily transported into a communion with the source of all life. Such conscious refashioning of the Supreme Being at the very least protects belief from the charge of naïve anthropomorphism while it maintains a sense of absolute power and mystery.

The idea of God as a life force or the source of all being sounds like a typical Generation Y answer to 'Do you believe in God?' In fact, The Spirit of Generation Y, the Australian sociological study published in 2007 but conducted over several years and based on 1272 respondents born after 1975, indicated a strong humanistic and scientific tendency in their spiritual beliefs. Yet the belief that divinity is the energy coursing through creation, including the human body, was the conclusion of the most innovative medical practitioner of the seventeenth century, Paracelsus, the man who was called in his time 'the Luther of medicine'. This honour referred to his reputation as a reformer of medicine, which, since the second century, had been hidebound by Galen's theories based on the four humours and four character types. Paracelsus was a contemporary of Luther and, and like the Protestant reformer, he had some annoying personal habits and character traits (also shared by Galen, for that matter), making them easy targets of the professional establishment. Nonetheless, Paracelsus—who counted among his friends and patients many well-known figures, including the great Dutch humanist Erasmus—was not only an effective healer; his profoundly religious bent furnished him with one of the great medical insights of all time: the power of the body to heal itself from many conditions and injuries. Paracelsus' discovery displayed some remarkable similarities to Eastern notions of divine energy, such as *Shakti* in India and the life force, Dao, in China. Self-healing processes of the body is also a principle feature of modern medical health, which seems obvious today but was by no means so in the past, when even simple wounds were burned with hot oil and skin gashes were filled with smoldering ashes and other volatile substances.

With such horrific treatments as the norm, the belief in the self-healing body would have many avid promoters for centuries to come, but perhaps its most notable ascendancy was in the nineteenth century, when the original New Age movement took hold in America's New England and spread throughout Europe and Australia.

New figures propounded extraordinary combinations of self-hypnosis, naturopathy, and the belief that deep within one's mind-body-spirit organism lies the holy grail of wellness and divinely ordained health. Such ideas, combining spirituality and health, found optimum conditions for growth at a time when the medical profession had not yet discovered vaccines or anti- bacterial cures for the many diseases that claimed young and old alike. (In some ways, the modern-day self-help and actualisation movement—dubbed SHAM by writer, essayist and investigative reporter Steve Salerno[4]— is the heir of the movement).

In fact, the boundaries between science and religion were not as clearly drawn in the nineteenth century as they are today, although there were official attempts underway to separate them.

The Christian Science of Mary Baker Eddy and the Anthroposophy of Rudolf Steiner were pre-eminent among the many spiritual movements that promoted their teachings as 'spiritual science'. Steiner, for example, asserted his views against 'the rigid materialists' who denied the mysterious realm, wherein the 'vital force' coursed through all living things just as the magnet produces the phenomena [sic] of attraction.[5] While he criticised scientific materialists, Steiner was nonetheless eager to use analogies drawn from the natural world. Yet, far from presenting a soundly reasoned argument, he also resorted to simple assertion; he declared, for example, that one's etheric body is a spiritual body which exists in another realm from the physical body and is not subject to the evidence-based investigation of science. How, then, is this etheric body perceived? Steiner answered by using a novel scientific notion drawn from Darwin: 'Spiritual science shows

---

4. Steve Salerno, *SHAM: Self-Help and Actualization Movement*, Nicholas Brealey Publishing, London & Boston, 2006.
5. Rudolf Steiner, 'The Emergence of the Etheric and Astral Bodies from their Envelopes', in *The Education of the Child*, (New York: Anthroposophic Press, 1994), 6, 5.

that human beings are capable of evolution, capable of bringing new worlds within their sphere by developing new organs of perception . . . For those who have developed the higher organs of perception, the etheric or life body is an object of perception and not merely an intellectual deduction.'[6] Evolution, other worlds and powerful organs all sound pretty concrete for a spiritual ideal, but that surely was the attraction of spiritual science.

During the rise of the science establishment the allure of its jargon to the spiritually minded founders of new movements is legendary, and might easily be explained with that old adage 'if you can't beat 'em, join 'em'. In their time, these movements became fully fledged religions with their own sets of beliefs, authoritative writings, educational systems and ritual practices.

Whatever motivated their initial promotion of the scientific value of their teachings, it seems to have gained its revolutionary character because of the vigorous moves to separate religion and science in the nineteenth century. As Peter Harrison has argued in a brilliant essay on science and religion, the power and promise of science as a purely secular and professional enterprise was a new phenomenon in the 1830s, with the emergence of the first professional bodies of scientists, such as the British Association for the Advancement of Science. Even the term 'scientist' was recent, having only been coined by William Whewell in 1833. By the mid-nineteenth century, the idea that the scientist was antagonistic to religion was being promoted by Thomas Huxley and his colleagues in the X-Club of naturalists and Darwin supporters, 'who sought with an evangelical fervor to establish a scientific status for natural history, to rid the discipline of women, amateurs, and parsons, and to place a secular science into the center of cultural life in Victorian England'.[7]

One could say that the people who were turfed out of the X-Club found a happy resting place in the halfway house of spiritual science. But like their hard-science half-brothers, they too were on a mission in the name of modernity, and it was to rid themselves of old-fashioned religion. Yet they fell between two stools, as it were, because they also

---

6. Steiner, op. cit., pp. 6, 7.
7. Peter Harrison, 'Science and Religion', in *Journal of Religion*, (January 2006): 81–106, 87, citing Ruth Barton, '"An Influential Set of Chaps": The X-Club and Royal Society Politics, 1864–65', in *British Journal for the History of Science*, 23 (1990): 53–81.

wished to create a universal religious philosophy which was impregnable to scientific criticism. They did so by claiming their philosophies represented the apex of science, the Absolute Truth, which was the true meaning of spiritual science. On a prosaic level, the promoters of spiritual science promised their adherents the practical advantages of a system in which 'man may shape his own future destiny, and know for a certainty that he can live hereafter, if he only wills.'[8] The practical and even material emphases which were central to the nineteenth-century spiritual movements are today replicated in the contemporary promoters of the so-called spiritual technologies which claim to offer the practitioner supreme mental and physical functioning.

Among the most well known of these 'spiritual technologies' is Scientology, founded by L. Ron Hubbard in 1953, now one in a sea of imitators and competitors. Elusive powers and idyllic dreams have always inspired the spiritual search, and this was particularly so when the mysterious East had already whispered some of its secrets to the colonials and intrepid visitors from Europe in the nineteenth century. The East's attraction to Western seekers was undoubtedly the myriad ways in which divinity was imagined, but most especially as a personal quality that was entirely within the power of the individual to nurture as 'self realisation'. The impact of Eastern religious philosophy continues today, but largely in hidden or latent ways, since so many of the Buddhist and Hindu notions that were popularised by orientalists have become embedded in much of our popular psychology and spirituality, especially through Carl Jung and the American guru of mythology, Joseph Campbell. Even Theosophy, an esoteric spiritual movement begun in the last quarter of the nineteenth century, owes many of its key notions to Buddhism and Hinduism. Although numerically small and seemingly inconsequential today, Theosophy's influence on so many writers and freethinkers was considerable and was the model for New Age eclecticism.

One of humankind's most curious obsessions is searching for paradise. Whether it is believed to be our original habitat or the final resting place for the spiritually evolved, imagining its bounty and perfection has been one of the abiding preoccupations of cultures the world over. Shambhala is the Eastern version of paradise, tucked

---

8. A Master's Letter, U.L.T. Pamphlet No 29, Bombay, Theosophy Company (India) Ltd, 24 April 1934, 10.

away behind the beyond in the Himalayas, inaccessible to the poor in spirit. In Europe's eastern reaches, the legend of the invisible city of Kitezh, immortalised in the 1907 opera of that name by the Russian composer Nikolai Rimsky-Korsakov, has similar spiritual significance. In the story, Kitezh, a city of churches, was attacked by marauding Mongols, and miraculously sank beneath Lake Svetloyar, protecting the pious inhabitants. According to legend, only the most pure in heart and soul can still hear the prayers and see the lights of the religious processions of this 'Russian Atlantis'.

In the West, a celestial paradise has been the traditional reward awaiting believers. Yet the temptation to invert the order of things, and see the perfect society not as a heavenly consequence of a life of righteousness but as its earthly breeding ground, has also been great. The most influential of all such 'experiments' was that described by Thomas More, in his early sixteenth-century tract *Utopia*. Since then the overwhelming desire to re-create More's ideal society, replete with egalitarian values and the idyllic qualities that delivered its inhabitants a deep sense of satisfaction, order and calm, has ignited the imagination of dreamers, practical men, and a few rogues.

Utopian dreams often founder on the very quality which they yearn to achieve, and that is order. In some ways, the Church's monastic orders, which were established on principles of poverty and obedience, as well as communal life, served as a template for many of the secular utopian experiments from the nineteenth century onward. Religious orders, after all, had a venerable history of subduing personalities from a variety of backgrounds and housing them under one roof, depriving them of the freedoms and the affections permitted ordinary folk, while pressing them into lifelong service to the Church. For some this was a happy arrangement, but for others it was a painful sentence, especially for women who were sealed off from society by a Church that was obsessed by the dangers to society posed by unmarried women. Curious as it may seem today, that obsession with female sexuality prompted bizarre solutions, from constrictive garb to high-walled enclosures. Some women fought against the restrictions imposed on them with daring and imagination, and occasionally managed to indulge themselves in the very pleasures—such as sex and fine clothes—that they were denied by the Church. Today, many women in religious orders have overturned the draconian restrictions that once ruled their lives, and their astonishing achievements and proud sense of womanhood have burst through the cocoon that had enclosed them in the past. And that is the story which begins this book's voyage of discovery.

Despite the passage of time, the curious obsessions of the past have hardly been exhausted even when events have overturned them and our ever-expanding knowledge has disproved some of their more outlandish claims. All the themes in this book have a continuing and lively currency today, whether it is magic or utopias, lost civilisations or the secret to health and wellbeing. If the old adage that there is nothing new under the sun is as true today as it was when Ecclesiastes wrote those words in antiquity, then we can be sure that the curious obsessions of today will keep us going well into the future.

Chapter 1
Nuns, Sex and The Vatican

In the international waters of the St Lawrence Seaway between Canada and the United States, nine women in white cassocks lay prostrate in a boat during the ceremony that would ordain them as priests and deacons. It was 25 July 2005, and under the auspices of Roman Catholic Women priests, an American organisation, they defied the Vatican's refusal to ordain women.[1] A similar ceremony was held in 2002 on the Danube River by the German group Weiheämter für Frauen between Germany and Austria. After ordination, these women ministered in a variety of settings but without the approval of the Vatican, although it seems they did find quiet support among some priests. While nuns are among those who wish to be ordained, they could not openly partake in such a brazen rejection of Vatican authority without endangering their own position in the Church and that of their order. But it was not for want of trying. Back in 1971, thirty-five-year-old Celine Goess—a Sister of Mercy of Holy Cross in Birch Run, Michigan—wrote to her bishop asking to be ordained as a deacon. She received a scathing letter in reply, telling her not to ask 'stupid' questions. Now, older but no less determined to see her desire fulfilled—if not for herself, then for others in the future—she serves on the board of the Detroit-based Women's Ordination Conference.[2]

Some of the most independent and spirited women I have ever met are Catholic nuns. They are also among the most educated and far-thinking about their faith. In fact, most of the nuns I've met are nothing like their popular image as old- fashioned and submissive. It is an image that dies hard, however, even among journalists, who

---

1. See https://romancatholicwomenpriests.org/
2. *National Catholic Reporter* (27 January 2006): 5–6.

are usually heat-seeking missiles when it comes to obliterating holier-than-thou religion. A story about the transformation of a Brigidine convent into a nursing care facility in Sydney was a case in point. All the nuns, including a sharp-minded octogenarian, wore slacks and T-shirts, but the press article featured a photo of the only nun who insisted on wearing the outmoded black-and-white habit, complete with wimple. It is just the sort of cliché that sells newspapers, but the reality of a nun's life is very different from the anachronistic stereotype the picture conveyed. In fact, journalists are none the wiser because they almost never ask nuns what they think of the Church. Quite possibly, they could not publish what they heard, because although nuns are often very critical of the hierarchy under which they nominally must take their place, the price of their hard-won freedoms is discretion, with 'off the record' comments being the norm. So while nuns do have a record of dissenting from authority, they are practised in keeping just enough distance to avoid arousing suspicion from or causing embarrassment to the Church, which they love and serve. Most nuns are not attention-seekers, and are used to achieving their ends within the system. Occasionally, however, feisty nuns who have risked all to follow their passion have failed to keep their superiors at bay. Occasionally they even have won them over.

Nuns have a long history of not running with the crowd, and some of them achieved fame well beyond their community. Today the popularity of German mystic, musician and abbess Hildegard von Bingen (1098–1179), the English Benedictine anchoress and mystic Julian of Norwich (1342–1423) and the Spanish Carmelite, Teresa of Avila (1515–82), has reached into all manner of contemporary spirituality. Their unfettered expressions of love for God and creation, in contrast to the turgid language of Church doctrine, have great appeal. Their love of beauty touches that part of the soul which recognises the glory of God in the wonders of creation. Even the conventions of monastic dress, already well established by the medieval period, were thrown aside by the Abbess Hildegard, who had no taste for rough cassocks and dark colours. In the illuminated manuscript *The Garden of Delights,* she depicted her nuns in the brilliant red and purple veils of secular dress. She was criticised at the time for her lavish tastes, but that did not stop her from designing elaborate crowns for her nuns.[3]

---

3. Elizabeth Kuhns, *The Habit: A History of the Clothing of Catholic Nuns* (New York: Doubleday, 2003).

Just about every country has had its share of remarkable nuns, but in Australia their daring had less to do with radical mysticism and extravagant visions and more to do with practical ventures, like Mary MacKillop's schools for deprived children. On reflection, perhaps the mystical union with God and the practical care of the needy are not so different. After all, it is in the act of humbly serving others that Jesus of the Gospels demonstrated to his followers the way of the Lord. Such active faith is certainly a more convincing act of holiness than the artificially contrived suffering and sacrifice that nuns were meant to endure behind the high walls of a convent or in an anchor hold. In any case, the realities of monastic life were not always consistent with that contemplative and holy ideal. A look at sixteenth-century Venice, which had more convents than anywhere in Europe, shows that a nun's life was not always as straitlaced as might be imagined.

There had been nuns in Venice since 640, when a Benedictine community of San Giovanni Evangelista (St John the Evangelist) was founded on the island of Torcello. A millennium later the population of nuns had grown to over three thousand, housed in fifty nunneries in Venice and its islands. By then, however, the reputation of nuns had suffered serious damage. The Protestant Reformation aimed some of its criticism of the Roman Church directly at the convents, which were described as dens of decadence, cruelty and lascivious behaviour. While this could not be said of all convents, neither could the published accounts of 'escapees' be dismissed as mere Protestant propaganda, since the Church itself was becoming sufficiently nervous about the activities of nuns to initiate its own investigation. Its recorded interviews with nuns have provided Oxford scholar Mary Laven with an unprecedented first-hand account of the cloistered life in the legendary city.[4]

Sending papal investigators into the Venetian nunneries to document the conditions therein was one of the means by which the Church of Rome in the late Renaissance arrived at its policy of enclosing nuns in communities behind high walls and secured by locked doors. The Council of Trent, the great reforming council of the Church, imposed policies that were meant to regulate the conduct of nuns, which at the time showed an alarming degree of variation and at times flagrant disregard for the life of chaste obedience

---

4. Mary Laven, *Virgins of Venice* (London: Viking, 2002).

expected of the Brides of Christ. This should have surprised no one, since convents were often considered 'dumping grounds' for girls and women of noble families who, for one reason or another, were not destined for married life. Perhaps the girl's parents had an insufficient dowry to pay for a good marriage or, conversely, there was no suitor worthy of her station. Alternatively, fear of childbirth, the single greatest cause of female mortality at the time, or unlovely looks, might send a young woman into the care of Mother Superior. The convent was also the only decent option for daughters born of illicit relationships, such as that between the world-renowned scientist Galileo Galilei and the Venetian beauty Marina Gamba. Illegitimate girls were unmarriageable except to Christ Himself.[5] A litany of practical rather than religious reasons could land a girl in a convent, but once there she was not necessarily on her own. Cousins, aunts and sisters might be there as well, and an extended family of women could take up residence in their 'corner' of the convent, complete with their own furnishings, wine store, and even the elegant clothes for which Venice was famous. This caused no end of rivalries in the communities in which they lived and undoubtedly led to the complaints against one another that the investigators duly noted. What is clear from Mary Laven's thorough study is that the nunneries experienced their greatest turmoil when friars and other men of the Church intervened in their affairs, often to oversee if not engineer the election of a favoured abbess. Sometimes these men of the cloth took up temporary residence in the convents just to partake of the nuns' larder and largesse. This frequent occurrence led to further scandalous rumours—especially when a nun and a friar went missing.

One of the most infamous of such cases in Italy involved the renowned Florentine Renaissance painter and Carmelite friar Filippo Lippi, whose love affair with the nun Lucrezia Buti resulted in two children. Eventually they were released from their vows and married; and Lippi continued to depict his wife in many of his religious paintings. The much-loved and -admired artist got off lightly; others who were caught *in flagrante delicto* and had no friends or relatives in the Vatican were threatened with imprisonment and castration. The most notable victim of the latter was the twelfth-century priest Peter Abelard, the poet and theologian who fell passionately in love

---

5. Dava Sobel, *Galileo's Daughter* (London: Fourth Estate, 1999).

with his student Heloise and made her pregnant. Heloise, herself a gifted woman of letters, bore their child, but refused 'the chains of marriage', and insisted instead on becoming a nun and a lifelong friend to Abelard. At a time when Church records show the practice of priests taking concubines was so widespread that ex-cathedra directives forbidding such priests from conducting mass were necessary, Abelard's castration at the behest of Heloise's uncle, Canon Fulbert, seems particularly tragic and cruel. Indeed, such became the decadence of the papacy later on, at the height of the Renaissance, the notorious Rodrigo Borgia, who became Pope Alexander VI in 1492, was even said to have fathered a son to his daughter Lucrezia. After him, Pope Leo X (Giovanni de Medici of Florence), who schemed to make his bastard cousin the next pope, wrote to his brother, saying, 'God has given us the papacy; let us enjoy it'. This they did, indulging in all the physical pleasures normally enjoyed by princes.[6]

It is no wonder that nuns were not always models of obedience and prayerful devotion, especially when some of them had no calling to the religious life. 'Forced vocations' were both a necessity and a curse. Although the Council of Trent made it clear that they were forbidden, and various measures were advocated to relax the stringent obligations of monastic life for women—such as fasting and wearing rough clothes—there were still those who argued in favour of forced vocations. In 1619 the Venetian Patriarch Giovanni Tiepolo warned:

> If the two thousand or more noblewomen, who in this City live locked up in convents as if in a public whorehouse, had been able or had wanted to dispose of themselves differently, what confusion! What damage! What disorder! What dangers! What scandals and what terrible consequences would have been witness for their families and for the City![7]

It would be hard to imagine women meekly accepting such statements without responding in kind. The Patriarch would later find his match in a nun who made it clear in her writings that her parents had forced her into the convent, and that she was but one of many women who were similarly deprived of their liberty. Elena Cassandra Tarabotti

---

6. William Manchester, *A World Lit Only by Fire: The Medieval Mind and the Renaissance: Portrait of an Age* (Boston, Ma: Back Bay Books, 1993), 38.

7. Laven, *Virgins of Venice*, 29.

(1604–52) was one of five daughters of a Venetian merchant family. She had been sent to a Benedictine monastery at the age of twelve, where she later took vows. Ironically, when the Patriarch Giovanni Tiepolo relaxed some of the restrictions on monastic life in 1629, she criticised his grand gesture by saying, essentially, that it was too little too late. Although she could not leave the monastery, she was permitted visitors, including members of the Accademia degli Ingoniti, progressive thinkers who kept her informed of current debate and brought her books. It was through them that her works were published. Initially she wrote two volumes, *Paternal Tyranny* (*La tirannia paterna*), and *The Monastic Hell* (*L'inferno monacale*), which she circulated to friends. Undoubtedly, such written criticism of the Church put her in some danger, and in 1643 she published a work in praise of monastic life, *The Monastic Paradise* (*Il paradiso monacale*), for those who had freely chosen it. In 1654, two years after her death, *Simplicity Deceived* (*La semplicita ingannata*) was published, reviving her criticism of the Church for allowing the coercion of women. Tarabotti, whose religious name was Arcangela, was by all accounts a woman ready to engage the debates of her day; she even penned a treatise on a woman's right to beautify herself.

The irony is that locking up women in convents only served to powerfully increase their allure, and the analogy to the whorehouse is easily made. While the median age of Catholic nuns today is around sixty, such was not the case in the past. A collection of virginal women, who in some instances were kept against their will, was shrouded in mystery yet dependent on the outside world for support. They earned their keep through the sale of baked goods and other manufactured products, such as textiles and lace. Such a 'female factory' drew its share of loitering males, who sometimes serenaded the nuns from the streets when they were not trying to get inside in the guise of handymen, delivery boys or even men of the cloth. Mary Laven describes an ever-widening net of security measures, from the old nuns keeping watch on the doors and window latches, to the patriarchs and magistrates who imposed punishments for even the most minor transgressions.[8]

However, these authorities were not foolproof, especially when it came to priests who had both privileged access and significant power

---

8. Laven, *Virgins of Venice*, 93.

over the nuns, which they were not beneath exploiting. In sixteenth-century Venice, Father Giovanni Pietro Lion, the confessor of 400 nuns of Convertite, was undoubtedly the most audacious of them all. Convertite was an unusual convent; founded to accommodate repentant prostitutes, it was filled with young and 'experienced' women. It was a perfect playground for the confessor, who regularly selected his conquest during the confession, made advances to her and, if she resisted, would threaten her with prison and beatings, which he administered personally. Behind the high walls, the nuns were 'protected' from the outside, but so too was Lion and his wily ways, which included parading the nuns naked (so that he could choose the best-looking), having them at will, and also taking their worldly goods for himself. In short, Convertite was Giovanni Pietro Lion's private bordello. In 1561, after the priest's antics were finally revealed in all their dissolute details to Ippolito Capilupi, the papal nuncio, Lion was led to the scaffold in Piazza San Marco and beheaded.

At a time when many priests had their own concubines, Lion's menagerie of nuns constituted one of the more extreme cases of sexual misconduct. But it should be said that sexual predation cannot be eliminated in the hothouse atmosphere of a monastery or convent where men and women are locked up in their youth and prevented from having the normal intimacies that bring pleasure and procreation. Indoctrination cannot entirely eliminate the powerful urges that course naturally through men and women, which makes the celibate life a continual challenge long after the vows are taken. And nuns, as much as priests, struggle to remain true to the ideal of chastity, which is shrouded in spiritual purity. This helps to explain the huge importance of the nun's habit, which for centuries hid the woman's body from herself as much as from others. 'The tongue talks of chastity, but the body reveals incontinence,' warned the early Church Father, St Jerome (340–420).[9]

This is not to suggest that women were disinclined to group together as celibates and lead a life of self-deprivation and devotion to God. Female Christian ascetics were living all around the Middle East by the third century. Spending their days in prayer, work and penitential exercises, women voluntarily adopted simple rough

---

9. Kuhns, *The Habit: A History of the Clothing of Catholic Nuns*, 70.

clothing as a way of marking themselves out as exceptionally committed to the religious life. Eventually, the distinctive garb of both men and women was viewed as a 'second baptism' and 'taking the habit' became synonymous with adopting a religious vocation.[10] But although the notion that 'cleanliness is next to godliness' is widely accepted among Christians today, the reverse was true for the first women monastics. In the fourth century, Paula (347–420), the abbess of a convent in Bethlehem, counselled a very different approach to personal hygiene. She followed the rule that had been laid down for women by the Egyptian hermit Pachomius (292–356), who is considered the founder of the first monastery. Paula, who was a great friend of St Jerome, dressed her nuns in identical garb, but warned them that 'a clean body and a clean dress mean an unclean soul'. She prohibited bathing and other kinds of personal care which were prevalent in both Roman and Jewish culture and ceremony. Whether Jerome approved of this rather extreme sign of piety is not known, but he did write frequently to his women friends extolling their decision to throw off personal adornment, jewellery and soft clothing in order to don the common tunic and cloak of the religious life.

The nun's habit underwent many changes, reflecting the particular region, period and work that nuns engaged in, but dark colours, such as black or brown, were prevalent. Occasionally it was white, especially for the 'betrothal' ceremony when the novice took vows to become a Bride of Christ. Every aspect of the habit was symbolic—even the act of getting dressed each morning was a prayerful ritual itself. Not surprisingly, the nun's habit, particularly when it belonged to an abbess or some other highly esteemed nun, was considered a holy relic, and upon her death the cloth was cut into pieces and distributed among the pious. A relic's miraculous power to heal the sick was the reason for many a nun's posthumous elevation to sainthood. Relics were a good source of income but with the advent of photography the Church witnessed a miracle of immense proportions. It was possible to sell photos of a saint with the same potential to do wonders for the faithful. The first example of this was Bernadette of Lourdes (1844–79), the young peasant girl from the Basque country who had several visions of the Virgin Mary by an underground spring in Lourdes, in the south of France. Bernadette quickly became a celebrity of

---

10. Kuhns, *The Habit: A History of the Clothing of Catholic Nuns*, 65.

her order, yet her extreme reluctance to occupy the role and her prolonged illness and early death from tuberculosis did little to dent a brisk trade in her photo, which became as good as a relic and a handy little earner for the order ever since.

Left: From the archives of the Sisters of Bon Secours USA, the original habit (left) and a version from the 1960s.

Right: The original headdress of the Franciscan Sisters of Charity also had a unique style. No matter the order, every aspect of a nun's habit was symbolic.

It is hard to imagine press photos of today's nuns or snippets of their T-shirts and sweat pants treated in such a reverential fashion, and indeed most modern nuns would be embarrassed by any attempt to sanctify them. They consider themselves to be 'normal' women, and a far cry from their counterparts in the early Church, who were taught to 'become men' because females were considered inferior and lacking in a soul. By current standards, the unfortunate nuns who were compelled to transcend their female nature probably deserved to be treated as exceptional beings! Today's nuns, who dress in contemporary fashion and sometimes wear jewellery and make-up and have stylishly coiffed hair, are demonstrating to the world that their religiosity is not be judged by how much they deny their femininity or subject their bodies to painful mortification and regimens of self-denial. On the contrary, today the spiritual worth of a nun's life is more likely to be measured by her peers, as much as by secular society, according to the community work in which she engages, the scholarship she pursues, the theology she espouses and the social causes she promotes.

It could be argued that being denied ordination in the Roman Catholic Church has given nuns a particular advantage. Rather than relying on patronage and the Machiavellian politics of the Vatican, nuns have earned their credentials and made their contributions within society itself and, more recently, its secular institutions. If deprivation is the mother of invention, then it can be seen to have worked wonders by increasing the number of women religious who have become 'Sister Professors' as they take up teaching jobs in theological colleges and universities and respond to the broader spiritual trends and aspirations of laypeople. American nuns like Elizabeth A Johnson, CSB,[11] Joan Chittister, OSB,[12] Mary Boys, SHNJM,[13] and an Australian, Maryanne Confoy, RSC,[14] who lecture young students about faith, ethics and theology, are no less significant in the life of society than a parish priest who ministers to a congregation. Indeed, one could argue it is the Sister Professor who is making the greater impact, planting seeds in a demographic most at risk of being left spiritually fallow or walking away from the Church. The Sister Professor caters to a segment of society that is becoming increasingly ignorant of its own heritage. University courses reach

---

11. Elizabeth Johnson is a religious sister of the Congregation of St Joseph and Distinguished Professor Theology, Fordham University, NY. She is the author of *The Church Women Want: Catholic Women in Dialogue* (New York: Crossroad, 2002).

12. Joan Chittister, Prioress of Benedictine Sisters of Erie, Pa. is a columnist for the *National Catholic Reporter* and the author of over thirty books, including *Called to Question: A Spiritual Memoir* (London/New York: Sheed and Ward, 2004)

13. Mary Boys, Professor of Theology, Union Theological Seminary, New York, is the author of five books, including *Has God Only One Blessing? Judaism as a Source of Christian Self-understanding* (New York: Paulist Press, 2000).

14. Maryanne Confoy, lecturer at Jesuit Theological College, Melbourne, is the author of *Morris West: Literary Maverick* (Brisbane: Wiley, 2005).

a younger generation that is not attending church yet is interested in learning about religion in an atmosphere that allows free inquiry and debate. In fact, many professor nuns have contributed mightily to that debate, having been themselves beneficiaries of the greatest reformation that occurred in the Church in the last century, the Second Vatican Council, which sat from 1962 to 1965. It is Vatican II that encouraged nuns to seek university education and which led them in the direction of profound reassessment of the traditions of their own Church and their role in it. The result is that professor nuns are doing more than shoring up the faith—they are reassessing and critically reinterpreting it, so that the role of women in the early Church and even in Scripture is shown to be more prominent and more authoritative than was previously believed.

Reclaiming the tradition for women is an ongoing movement with many players, including nuns like Sister Christine Schenk, co-founder of FutureChurch,[15] an American organisation with international reach in six lanuages, that calls for women's equality in the Roman Catholic Church and celebrates 22 July, the feast day of Mary Magdalene, as a national observance. In Australia, the national organisation Women Church keeps the flame of feminism alive. Although it includes both lay women and nuns and is Catholic in origin, it prefers the small-c 'catholic' as a true description of its ecumenical vision. Women Church has been responsible for inviting women like Sister Joan Chittister to Australia. Of course, conservatives view organisations such as Women Church, with its preference for feminist scholar nuns, as far from positive. They see it as a sign of the loss of faith. Such attitudes compel Chittister, an American Benedictine nun, to encourage her audiences to demonstrate 'holy anger', by which she means a willingness to protest injustice, including the Church's oppression of women. But 'holy anger' has its costs, and not all nuns have been able to sustain it and remain in holy orders.

British nun Sister Lavinia Byrne, widely known for her segments on BBC Radio's 'Thought for the Day', was compelled to leave her order after more than thirty-five years when her book, *Women at the Altar*, which called for women's ordination, was ordered to be

---

15. https://www.futurechurch.org/women-in-church-leadership/women-witnesses-of-mercy/november-woman-witness-of-mercy-sister-christine Sr Christine Schenk CSJ lead the organisation from 1990–2013, and is a regular columnist for the National Catholic Reporter under the heading 'Simply Spirit'.

shredded by the Vatican.[16] Subsequently, her fame and reputation in the general community rose even further. Going where she is needed, Byrne took a job at an Anglican theological college, Westcott House in Cambridge, teaching women *en route* to ordination in the Anglican Church. When I asked her if training women priests of the Anglican Church caused her anguish, given her own Church's attitude towards her, she said there was one thing that hurt more. 'The thing that really does cause me pain, funnily enough, is the fact that I am invited to preach in Canterbury Cathedral, Westminster Abbey, Anglican churches all over the country, and college chapels in Oxford and Cambridge, so I have complete access to those major pulpits, but I am never invited to preach in a Roman Catholic church, because, as a woman, I can't.'[17]

While some nuns challenge the received tradition by stepping further out into secular society and embracing many of its humanist values, the daring has sometimes gone in the opposite direction. Rather than nuns yearning to merge with secular society, there are examples of women in society who have urged the Church to recognise the work they do in society as holy. So it was in the early 1900s with a young woman in Sydney who was determined to gain recognition from the Vatican for 'her women', nurses who ministered to the poor and the sick, yet broke with convention by bringing care into their homes. While domiciliary care was not unknown, it was rare, and the women who lived together under the tutelage and spiritual guidance of Eileen O'Connor scandalised polite Sydney society and alarmed the Roman Catholic Church. Entering private homes unchaperoned, these nursing sisters were called loose women, and worse, they were shunned in public and were even spat on. Meanwhile, their inspirational leader, the wheelchair-bound Eileen O'Connor, who believed she was led into this work by the Virgin Mary, struggled to persuade bishops and the Pope himself to consecrate them as an order. Together with her friend and supporter Father Edward McGrath, Eileen was both founder and 'mother' of a de facto order of sisters in Sydney, Our Lady's Nurses for the Poor.

It is hard to imagine that a frail girl, confined to her bed and a wheelchair for most of her short life, could become the object of

---

16. Lavinia Byrne, *Women at the Altar*, (London: Mowbray, 1994).
17. Interview, 'The Spirit of Things', ABC Radio National, 15 April 2001.

scurrilous rumours that bordered on the pornographic. But, moved by a profound faith in God, Eileen O'Connor was determined to transcend the obstacles put in her way and provide both physical and spiritual support to others, including the young Father McGrath, who was at odds with his Sacred Heart Order for helping to acquire property for Our Lady's Nurses. She even travelled to Rome in 1915 to state her case to Pope Benedict XV and also to request that Father McGrath not be punished by his order for helping her to realise her dream. By 1919, bedridden, in considerable ill-health, and with her friend Father McGrath in exile, Eileen still hadn't managed to gain the endorsement she craved for her Nurses, but she diligently wrote a rule for them should they receive the Vatican's blessing. In 1932, eleven years after her death, the blessing was finally given, along with a great deal of praise for her work. But as Bishop David Walker of the Diocese of Broken Bay states in John Hosie's masterful biography, *Eileen*, 'like many holy people in the history of the Church, Eileen came into conflict with the Leadership of the Church'.[18] In the case of Eileen O'Connor, a woman who advocated 'a group of religious plunge into the secular world of families, streets, professional people and governmental agencies',[19] that conflict was the result of the Church's resistance to the radical idea that nuns could lead a holy life while working outside the convent walls and in society. Her letters reveal that she was well aware that her proposal to minister to the needs of the slums was far removed from the idea of the convent, but she knew where the greater holiness lay. In that, she was very much closer to the example of her Lord than to the Church of her day.

Eileen O'Connor never experienced the fulfilment and joy of being recognised and supported by her Church, although it did not prevent her from working to establish an order of nursing sisters to the poor. Today, women are still showing just as much determination as Eileen, but almost a hundred years later the conditions are ripe for an even greater show of boldness, as evidenced by the Womenpriests movement. While most of the women who have been 'ordained' as Catholic priests are obliged to work discreetly in nursing homes and other care facilities, and cannot expect to work as parish priests, some have refused to 'hide their lights under a bushel'. In a parish

18. John Hosie, *Eileen: The Life of Eilen O'Connor* (Sydney: St Pauls, 2004).
19. Hosie, *Eileen*, 268.

in Rochester, New York, two women priests, Mary Ramerman and Denise Donato, with the support of a former diocesan priest, minister a 'full service' to a Catholic community. Begun on Valentine's Day 1999, Spiritus Christi's first mass had 1100 attendees. It holds weekend Masses plus daily Mass, enrols hundreds of children in faith formation classes and hosts many weddings, including LGBT unions, annually. The church originally reflected the family experience of both women priests, each of whom is married with three children. They created an inclusive church of equals which they believe is what people want.

Chapter 2
Dying to Know About the Stars

Will the universe ever end? Can we rely on it to continue on
and on? The most recent measurements of the velocities of
recession of very distant objects in the universe, supernovae,
which can serve as standard 'light beacons' at distances of
about 10 to 12 billion light years from us, indicate that the
universe is not only still expanding but that it is accelerating
in its expansion and will, unless we discover a breaking
mechanism, expand forever—an empirically infinite universe.[1]

So proclaimed Father George Coyne, the longest serving director of
the Vatican Observatory (1978–2006), based at Castel Gandolfo (the
Pope's residence) and in Arizona, in a speech given at the Vatican in
November 2005. It might seem like a contradiction to have a stargazer
in the heart of the Church of Rome; after all, we are accustomed to
hearing about Galileo, whose celestial observations nearly cost him
his life in 1633 at the hands of the Inquisition. But as his case and
many others both before and after his time make clear, stargazing
from within the Church is not as novel as it seems. In the annals of
Church history, many popes were fascinated by the movement of the
stars and the exploration of the unknown. Some even liked to have
their star charts secretly read by astrologers—but only if the prognosis
was not fatal. More of that later.

One of the great myths of the present is that religion always
stands in the way of scientific progress. The truth is that the Church
sometimes behaved with shocking indifference to human life and

---

1. George Coyne, 'Infinite Wonder of the Divine', in *The Tablet* (10 December
   2005): 6.

suffering in its efforts to suppress free thought. But the irony is that some of the most intellectually adventurous minds ever to push forward the frontiers of knowledge were men of the cloth. Among these colourful personalities, two of whom will occupy this chapter, are Giordano Bruno and Athanasius Kircher. Separated only by a generation, they were stargazers, inventors and philosophers who, four hundred years ago, knew what cosmologists today continually confirm in their exploration of space. Their experience demonstrates that the religious mind was not incapable or unwilling to explore the unknown; but if their discoveries were presented in a way that challenged the authority of the Church, then the Inquisition would stop at nothing to suppress them. In fact, before the Enlightenment gave us secular universities, the quest for knowledge was in the hands of priests and monks who often risked their lives to reveal new facts about the cosmos and human history.

This is largely forgotten today, when the swiftest way to discredit a public figure, especially a scientist, is to point out that he or she is a believing Christian. The assumption is that high-calibre intelligence, scientific thinking and rational acuity is less likely to be present in a person who believes in God. It is a view that lingers and occasionally is given a high profile by prominent scientists like Nobel Laureate Herbert A. Hauptman (Chemistry, 1985) who stated that belief in God is not only incompatible with science, but also 'damaging to the well-being of the human race'.[2] While many of his colleagues do not agree with him, they are often forced into silence by the intolerance of the scientific community. The director of the National Human Genome Research Institute, Francis S. Collins, admitted that speaking freely about his Christian faith is often taboo in scientific circles.[3] Occasionally, the discreet silence is overshadowed by the noisy creation evolution debate, in which groups like Answers in Genesis, which promote a literalist reading of the Bible, make a mockery of the brilliant and subtle religious minds that have paved our way to the scientific revolution. After all, Isaac Newton (1642–1727), who discovered gravity, wrote more about the Bible than the laws of nature, a fact that was undoubtedly ignored in a poll by the Royal Society, when Newton was voted a greater scientist than Einstein. (In fact,

---

2. *New York Times*, 23 August, 2005.
3. *New York Times*, 23 August, 2005.

both of them had a keen interest in the nature of God.) One of the key observations from the poll was that Newton led the transition from an era of superstition to one of modern scientific method.[4] It is tacitly accepted that superstition is not to be confused with Newton's faith in a great God, whose glory he found in the unfolding knowledge of the universe. Three hundred years later, Einstein appeared to echo Newton, when he declared that God was to be found in the laws of creation itself, specifically his theory of relativity. Clearly for Newton, as for Einstein, God's majesty and mystery contained as-yet-unknown and undiscovered truths about the universe which their scientific minds might reveal through investigation, calculation and speculation. This was the fundamental difference between the scientist's faith in God and the dogmatic formulas of the Church which, literally understood, give rise to superstition.

For some brilliant minds, however, the defeat of superstition at the hands of the scientific revolution did not come early enough. One of the most extraordinary men to have received the full measure of the Church's punishment was Giordano Bruno. His professed faith in God was not a barrier to his scientific discoveries but an invitation to conceive of divinity as something greater than the Church had previously imagined. For this he might have been regarded as a curiosity best ignored, but Bruno's tendency to attack in the most venomous and obscene language the superstitious aspects of Christian belief made him an especially visible target, despite his best efforts to evade the authorities.

In 1600 Giordano Bruno was marched through the streets of Rome as anything but a true believer in almighty God. Bound and gagged, with steel spikes driven through his chin to keep him from addressing the throngs who jeered at him, the short, bedraggled man, emaciated from eight years in prison, was paraded as a heretic destined for the fate that awaited all such men and women: death by burning at the stake. Despite Bruno's criticisms of Christian belief, he did count himself a Christian believer in God and was a man of uncommon vision and imagination. Although he lacked the usual social graces and calculated charm that subversive and ambitious clerics needed to survive, his arguments about the infinite vastness of the universe and the interconnectedness of all life leave one without any doubt that he died for truth itself.

---

4. *News in Science*, 24 November 2005.

Giordano Bruno was born in 1548 in Nola, a small town near Naples, which gave him his nickname, 'the Nolan'. In 1563, he entered the St Domenico Monastery, Naples, where he remained until 1576. In fact, his departure set the pattern for the rest of his life. He was forced to flee in secret after having aroused the ire of his fellow monks with his arguments against Aristotle's physical universe, around which so much Catholic doctrine was built. Aristotle had mapped out the universe in such a way that the heavens consisted of concentric spheres on which the planets, the sun and the moon rotated at different speeds with the earth at the centre and with the outermost sphere propelled by the Prime Mover. Ever since Thomas Aquinas (1225–1274) synthesised Christianity and Aristotelian philosophy in his massively detailed work *Summa Theologica*, for which he was canonised, it would become a serious offence to cast aspersions on Aristotle. Ironic as it may seem for the Church to be safeguarding the reputation of a Greek philosopher who knew no Christian God, it was the same Church that hunted down renegade monks for reading banned books on alchemy, astrology and mysticism.

The extraordinary renegade monk Giordano Bruno was able to reconcile his sixteenth-century scientific discoveries with his faith in God. His venomous attacks on the superstitions of the Church, however, elicited the full measure of the Church's censure.

It was not only for these arcane tastes, relished by Giordano Bruno, that he was denounced by a fellow monk, however, but also for his reading of the Christian humanist Desiderius Erasmus (1466–1536). This remarkable man was persona non-grata in the Church because he daringly argued for religious tolerance at the height of the religious wars between Protestants and Catholics. A much more savvy operator than Bruno would ever be, Erasmus was one of the most influential and outspoken monks of his time. A staunch advocate for the sanctity of individual conscience in matters of belief and a relentless critic of the barbaric practices of the Inquisition, he managed to elude its burning fires by travelling abroad—including to England—and finding protection among influential patrons. Bruno would attempt to do the same in the years after 1576, when he was tried and convicted, in absentia, of heresy. Unwilling to face his accusers, he was condemned to leading an itinerant life, an outlaw in a cassock who always had to stay ten steps ahead of his pursuers. This undoubtedly suited his temperament because, as much as he was on the run from the Church, he was also in constant pursuit of new knowledge, new contacts and new patrons. They were not hard to find, since the burgeoning interest in ancient Egyptian religion, Hebrew and Arabic texts, Greek philosophy, science and the occult was already well underway a century and a half before Bruno's time. The Renaissance sparked a massive search for ancient manuscripts, for which wealthy families such as the Medicis of Florence were able to pay.

While the region of Tuscany benefited from the powerful Medici family, who built up their libraries and patronised the arts, the northern port city of Venice was the crossroads for new ideas in everything from fashion to philosophy. Venice was a cosmopolitan city with a fiercely independent spirit. With its powerful shipping empire, its ties to the East and its astonishing prosperity, Venice could afford to conduct its affairs in a way unique in Europe at the time. Governed by a two-tiered *collegio* (consisting of an upper and a lower house) drawn from the wealthiest families, Venice went its own way on many matters, including religion, and was many times the object of Rome's censure, not least because Venice proved a fickle partner in the Holy League of 1495, having joined forces in 1513 with France in a war against Milan and the papacy. Venice had harboured a deep distrust of the Roman Inquisition—a system of tribunals set up to halt the spread of Protestantism—and had developed its own parallel institution. For centuries after Pope Gregory IX founded the Inquisition in 1231, Venetians had refused to comply with papal orders. As a seaside city greatly influenced by its trading empire, it was inclined to offer asylum to a host of unorthodox thinkers and heretics, including Protestants. In such an environment, the fugitive teacher and writer Giordano Bruno thought he would find a temporary haven.

One of the most important features of Venice for a man like Giordano Bruno was that its booksellers were permitted to trade in forbidden books, those that the Pope in Rome had proscribed. A simple deal had been struck between Pope Clement VIII and the governors of Venice. They agreed to build more churches if the Roman Inquisitors were kept away from the city. So it seemed to be a risk worth taking when, in 1591, after living in exile in London, Paris and Germany, Bruno accepted an invitation from a wealthy Venetian senator, Giovanni Mocenigo. The senator wished to be tutored in one of Bruno's pet subjects, the art of memory. In fact, mnemonic techniques were widely practised in occult circles (which Bruno frequented) in order to commit to memory secret formulas and theories that could not be written down. It was an ancient art, and Bruno's contribution, which he claimed was a rational process and not one that relied on sorcery, would be his last published work, *On the Composition of Images, Signs and Ideas* (1591). Mocenigo proved to be a demanding host, and possibly in the pay of the Venetian Inquisition as well. He had Bruno arrested just as he was planning to leave the city. It would be the beginning of the end for the Nolan whose heretical views and magical practices had made him a notorious figure to the Church.

Although Bruno was suspected of sorcery and witchcraft when he was interrogated about the purpose of his mnemonics, it was his philosophical and theological ideas that posed a far greater threat to the Venetian Inquisitors. In effect, Bruno believed that God was infinite and the creator of a universe that was infinite, containing not one world but numberless worlds. *On the Infinite Universe and Worlds* (1584) contends that all the stars are suns with their own planets, inhabited by intelligent beings. It was one thing to call into question the singularity of the earth and humanity as recounted in the Book of Genesis; it was quite another to propose an indeterminate God, who bore no similarity to the Father God of the Christian Church, who, as the Christian Bible recounts, had a unique and exclusive relationship with humankind through his one and only son. It is impossible to overstate the incompatibility of Bruno's infinite God with the central Christian belief in God's son, whose life, death and resurrection offered the only salvation for humanity. For without its unique promise of salvation the Christian Church had no reason to exist.

Bruno's God, on the contrary, was incapable of being boxed into a particular set of characteristics or a singular story of salvation, and furthermore Bruno believed (along with some other heretics) that the idea of the Trinity was merely an invention of the early Church and hammered into doctrine during the political machinations of the Emperor Constantine at the First Council of Nicaea in 325. (It was then that the hodge- podge of Christian beliefs were sifted, sorted and selected to form the doctrinal basis of the Church.) Nor did Bruno's infinite God fit into the Biblical time frame of Creation, Salvation and the Last Judgement. A man with a voracious appetite for knowledge, Bruno found a more compelling source of beliefs in Egyptian religion, which was then gaining popularity among the readers of the banned *Corpus Hermeticum*, a newly discovered work that was attributed to the philosopher Hermes Trismegistus. Although Trismegistus was believed to have lived in the earliest period of ancient Egypt, scholars today ascribe to the *Hermeticum* a much later date, probably around the fourth century.

Bruno may have looked like he was switching his allegiances, but his investigations into 'the hermetic tradition' based on Hermes Trismegistus led him to believe that Egyptian religion was where the real roots of Christianity were to be found. He even argued that the Egyptian cross, the ankh, with its distinctive circle or 'head' at the top, was more powerful than the Roman cross. It was just the sort of dig at the Church which Bruno would have delighted in making, but it was also among his more irrelevant observations. If anything, Bruno was far more interesting when he was not taking pot shots at the Church about its alleged irrationality and its pedantries, but was stretching the boundaries of scientific knowledge with his cosmological speculations. When he dared to imagine unbounded space and a universe where all matter is related to all other matter, and when he observed that human beings were continuous with every other living thing in the universe, he was positively inspirational. How could he have known that all life possesses the building block of carbon? This is from his satirical play *The Chandler* (*Il Condeai*; candle-maker or torchbearer), written in Paris and published in 1582, in which the chandler who advertises his wares on the streets is obviously a mouthpiece for Bruno himself:

> Behold in the candle born by this chandler, to whom I give birth, that which shall clarify certain shadows of ideas . . . I need not instruct you of my belief. Time gives all and takes all away; everything changes but nothing perishes. One only is immutable, eternal and ever endures, one and the same with itself. With this philosophy my spirit grows, my mind expands. Whereof however obscure the night may be I await the daybreak, and they who dwell in day look for night. Rejoice therefore, and keep whole if you can and return love for love.[5]

When he says 'One only is immutable, eternal and ever endures', Bruno is describing his God. It is not the divine personality etched in the Bible, but the abstract notion of the life force, the light source and the overwhelming love, which is how God is described by the mystics. In particular, Bruno was familiar with the mystical Jewish writings, known collectively as the Kabbalah (the 'received tradition'), which became so popular in the Renaissance that Christians adopted their own version called Christian Cabala. The radical nature of the Kabbalah derives from its claim to be an account of the creation before the creation as written in Genesis. In a nutshell, it is a story of God's overwhelming love of creation, which is reciprocated by creation's intense desire to return to the Godhead, which is represented by light. The most important part of the story is that every bit of creation contains a divine spark, which originally came from the Light Source and wishes to reunite with it. One can easily see here shades of the Egyptian Sun God, considered to be the first manifestation of monotheism before Abraham, but it is also a perfect poetic analogy for the carbon that inheres in everything. It makes abundantly clear that there is a bit of God, the life giver, in all of us.

---

5. John J Kessler, *Giordano Bruno: The Forgotten Philosopher*, www.positiveatheism. org/hist/bruno.htm

Keen to learn as much as he could, Bruno absorbed ideas wherever he went, from France and Switzerland to England and Germany. As with his science so with his radical theology. It is not hard to hear the echoes of those who went before him. In Germany, for instance, the legacy of Johannes Reuchlin (1455–1522), the son of a Dominican official, encouraged the study in universities of Hebrew books. Reuchlin's own preference was for the mystical writings of the the Jewish Kabbalah, which seemed to provide the theosophical language for reconciling science and mystical faith, undoubtedly the very quality of the Kabbalah that Bruno himself appreciated. Reuchlin published his ideas in *De Arte Cabbalistica* in 1517 (also known as *The Science of the Kabbalah*). By then Reuchlin had managed to save the Hebrew books of Germany from going up in flames when, undoubtedly for his own purposes rather than as a sympathetic gesture toward the Jews, he advised Emperor Maxmilian I against it. The Germany that Bruno visited also would have benefited from Reuchlin's groundbreaking Hebrew grammar and lexicon, *De Rudimentis Hebraicis*, published in 1506, a volume that Bruno might have wished to get his hands on to brush up his own rudimentary knowledge of the language.

Although historians, such as Daniel Matt, editor of the Pritzker edition of *The Zohar,* usually trace the Kabbalah's origins to thirteenth-century Spain, it is possible that its mystical motifs are universal, and also can be discerned in Medieval Christian mysticism. In Germany Bruno would have encountered the ideas of Meister Eckhart (1260–1327/8), the distinguished Dominican preacher, provincial of his order and mystic who was both revered in his time yet subject to the Roman Inquisition in the last years of his life. Eckhart's radical teachings, which are popular today, at times seem utterly bizarre. In fact, they are the non-rational, nonsensical and paradoxical utterances of a mystic whose experience of the divine is imagined as the melding of the human and the divine into one. While popular mysticism can cover a multitude of sins, such as the inflation of the self to divine proportions, it is also true that the language of mystical union never conforms to conventional Christian beliefs. Among Eckhart's most consistent messages, however, was that God was in man and man was in God, and that it was possible to attain such a profound union with God that one could see the world through His eyes and achieve a state of sanctification. In many of his sayings, Eckhart rejected the importance of overt works and public piety and seemed to subvert the role of the priest as intermediary between God and man. It was a view that was in keeping with Giordano Bruno's outlook 350 years later, although they arrived at their conclusions by different means. The unitary state for Eckhart was entered into by the mind and heart of the faithful, whereas for Bruno the unitary state was arrived at through lifelong study, which would uncover the true nature of things, that God and the natural world are one.

It is worth noting, lest Bruno's tragic fate result in an overly romantic portrait of the man, that his entire approach to Christian beliefs—at a time when the Church was employing rationalistic arguments to develop doctrine and canon law—was to ridicule and debunk them. One can have some sympathy for Bruno, for whom there was no comfortable scientific establishment to offer support as he struggled to make his point with the Church. Indeed, he endured a gruelling eight years in jail in the most cruel physical conditions while he strenuously argued his case. Yet the circumstances of his time cast him into the role of a desperado. As a new breed of scientific sleuth, beholden to no one but himself, Bruno was bound to see himself up against a Rome that was intent on subverting the sciences and philosophy to serve Christian beliefs at the expense of a deeper understanding of God. To an extent he was correct. Even Thomas Aquinas is said to have cast doubt on the worth of his edifice of theological reasoning, when near the end of his life a mystical experience, an epiphany, delivered him the most profound understanding of his faith that all the numbered paragraphs in his great work, the *Summa Theologica*, could not. The true role of religion, after all, is to inculcate faith, humility and devotion to the mystery that is at the core of existence. When both the Church and visionaries like Giordano Bruno lost sight of this and instead became arrogant, pontificating 'know-it-alls' they were set on a collision course.

For his part, Bruno was a believing Christian, just as Meister Eckhart was one of the most revered preachers of his day. Some of their views were remarkably similar, but Eckhart expressed them in mystical terms which allowed him some poetic licence. Three hundred and fifty years later, when Bruno rejected the Catholic doctrine of the Eucharist, which holds that the bread and the wine consecrated by a priest literally become the body and blood of Christ, he not only proposed that the Eucharist should be considered a *symbolic* rite, a view held by the Protestants of his day, he also declared it to be a position which did not contradict the natural world and scientific truth. Similarly, Meister Eckhart had made statements about the nature of God which would have been equally shocking to Catholic ears: 'God has no image nor likeness of himself, seeing that he is intrinsically all good, truth and being'; 'God needs no image nor has no image; without image, likeness or means does God work in the soul'; 'Were any image present there would not be real union and in real union lies your whole beatitude.'[6] With these and many other similar statements Eckhart, long before Bruno, intuited that the real significance of God, His actual working in one's life and soul, cannot be contained in or exclusively represented by the artefacts of Christian piety—such as statues, paintings, talismans and even the sacraments—but is to be found in a holy union that transcends those things, which is wholly spiritual and expressed through a blessed life. Despite Eckhart's high standing in the Church, late in life he was called before the Archbishop of Cologne and tried for heresy in 1326. Fortunately for Eckhart, he escaped the final humiliation of Pope John XXII's final judgement in 1329, for he died before some of his teachings were condemned as heretical. Even the Inquisition, which confined itself to his teachings and not the man, realised that the Dominican 'Meister' himself was beyond reproach.

It has been suggested that Eckhart would have been left alone were it not for the fact that the Archbishop of Cologne was a Franciscan, an order that was in conflict with the Dominicans at the time. Either way, the politics of dissent should always be kept in mind, since priests were not unaware of the dangers of openly expressing their views and they were better than most at escaping censorship and the depredations of the Inquisition, which, it should be said, handled them with far more consideration and latitude than it did laypeople. It also helped immeasurably to have an old friend in a pope or an archbishop, who might be disposed to protecting a daring theologian or a bold scientist. Such was the case for the German Jesuit, Athanasius Kircher (1601/2–80), whose brilliant knowledge of languages, mechanics, cosmology, mathematics, ancient history, music, medicine, the occult and foreign cultures has earned him the reputation as the German Leonardo da Vinci. With such a voracious appetite for knowledge

6. Ursula Fleming, *Meister Eckhart: The Man From Whom God Hid Nothing*, (New York: Harper Collins, 1988), 91, 92–93.

and an original mind (he invented many devices which are held in the Kircher Museum in Rome), he was bound to get on the wrong side of some established orthodoxies.

The fundamental belief that the earth was fixed at the centre of the universe, which was limited and not infinite, was one such orthodoxy. In 1530, after thirty years of private study, the Polish cleric and astronomer Nicolas Copernicus (1473–1543) completed *De Revolutionibus*, which asserted that the earth rotated on its axis once daily and travelled around the sun once yearly. Although some parts of his work were circulated among friends, it was not published, and Copernicus would never know the uproar his work caused; it was for later scholars like Galileo (1564–1642) and Bruno, who accepted and promoted Copernicus's discovery, to suffer the consequences.

Athanasius Kircher, however, based in Rome from the mid-1630s and teaching mathematics at the Collegio Romano, did manage to escape prosecution despite holding views entirely consistent with all three of his forerunners. It seems that, like Copernicus, Kircher did not rush to publish a controversial work on cosmology, and was

Galileo's secular astronomical investigations challenged the authority of the Church and its portrayal of heaven and earth.

more circumspect than others in his midst, including his student and collaborator Father Kaspar Schott, about expressing his contentious views on astronomy. Yet he had a good friend in Pope Alexander VII (1655–67), who was a renowned scholar himself and inclined to mix with a disparate range of friends, including freethinking Protestants, as did Kircher himself. According to the Renaissance scholar Ingrid Rowland, Alexander VII's papacy was an optimistic time for scholarly endeavour, and it was then that Kircher decided to commit his cosmological speculations to print.[7]

Unlike Galileo, who was secular and therefore represented a serious challenge to the clerical class, and Bruno, who was a renegade monk, Kircher, the brilliant polymath, Jesuit and teacher in the heart of Rome, was not an easy target for censorship. It also helped that he wrote his treatise in what might be called a pure science fiction or fantasy genre. His *Ecstatic Celestial Journey* (*Itinerarium Extaticum s, opificium coeleste*) is the account of a dream in which a beautiful celestial being, Cosmiel, offers to take Theodidactus ('taught by God') through the secret realms of Heaven and Earth. Not that dream-like literature was anything new; the Roman statesman and poet Cicero (106–43 BC) had also committed his cosmic travels to a dream essay in *Somnium Scipionis*, and in 1321 Dante Alighieri produced one of the greatest works of Western literature, *The Divine Comedy*, in which the central figure is led in a dream through the realms of Hell, Purgatory and Paradise, accompanied by his muse and love, Beatrice. In Athanasius' essay, the subject matter is decidedly cosmic, not moral, and Theodidactus' questions to Cosmiel are answered with dry wit, a sharp tongue and a scientist's wisdom. After casting doubt on Aristotle's knowledge of nature, she says, '. . . we angels observe that human thoughts, unless they are based on experiments, often wander as far from the truth as the earth is distant from the moon.'[8] However, it is when Theodidactus travels beyond Saturn with his celestial companion that he makes the most radical discovery. On learning that he is expecting to 'bump into' the fixed stars, Cosmiel takes him to task for his naivety:

---

7. Ingrid Rowland, *Ecstatic Journey: Athanasius Kircher in Baroque Rome* (Chicago: University of Chicago Press, 2000).
8. Quoted in Rowland, *Ecstatic Journey: Athanasius Kircher in Baroque Rome*, 194.

My Theodidactus, now I truly see that you are excessively simple of mind, and more gullible than average when it comes to believing anyone else's opinion. The crystalline sphere you are looking for cannot be found in nature, and there is no basis for the idea that the stars are fixed on such a sphere. Look around, examine everything around you, wander the whole Universe, and you will find nothing but the clear, light, subtle breeze of the great ethereal Ocean, enclosed by no boundaries [my emphasis], that you perceive all around us.[9]

Here we find Giordano Bruno's infinite universe reiterated thirty years later in plain language in the work of Athanasius Kircher. There were other similarities to Bruno's ideas embedded in Kircher's work as well. Just as Bruno posited a World Soul, out of which infinite possibilities emanated, so Kircher wrote of 'the supreme Archetypal Mind . . . so full of ideas for possible things [that] he wanted to establish this universe . . . with a numberless variety of spheres'.[10] But why should there be an infinite number of spheres? Bruno reportedly had said that the answer lay in the fact that 'God needed the world as much as the world needed God and that *God would be nothing without the world*, and for this reason God did nothing but create new worlds'.

The idea that the all-powerful God of the Bible was somehow dependent on the natural world was regarded as heretical by the Catholic Church. Yet it is central to the Jewish Kabbalah, the mystical writings that imagine God's creation of the world motivated entirely by his overwhelming love for it. Both Bruno and Kircher were familiar with the Kabbalah (Kircher having learnt Hebrew at a young age from a rabbi in Meinz just as Johannes Reuchlin had done before him), and may well have been inspired by its most famous work, the Zohar, the Book of Radiance, which opens with an account of the creation *before* the creation recorded in Genesis. It is a breathtaking narrative which contemporary thinkers have likened to accounts of the Big Bang, when the universe was unleashed from a void of timeless darkness. In any case, Christian interest in the Kabbalah would never be greater than during the lifetimes of Bruno and Kircher, who benefited from

---

9. Quoted in Rowland, *Ecstatic Journey: Athanasius Kircher in Baroque Rome,* 196.
10. Rowland, *Ecstatic Journey: Athanasius Kircher in Baroque Rome,* 197. Moshe Cordovero, 16th Century Kabbalist living in Sephad, Israel, also explains God in a Universe of endless spheres.

a previous generation's translations and studies of it, not only by Reuchlin in Germany, but Giovanni Pico della Mirandola (1463–94) in Italy.

Despite his bold statements about the nature of God and the universe, Kircher was not summoned before the Inquisition, nor was he removed from his post, but the Jesuit censors did lodge a complaint against *The Ecstatic Celestial Journey*'s notion of an infinite universe. It went no further, since Kircher did, after all, insert an ingenious disclaimer in the text. When Theodidactus' eyes were newly opened, Cosmiel told him that while the immense universe only appeared to be boundless, it was perfectly finite in the eyes of God! The anarchic notion of infinite space that contradicts the very meaning of the word 'cosmos', which connotes an ordered universe, is put to rest, and God's ultimate superiority over the whole universe is restored in the blink of an eye. Fortunately, Kircher's loyal friend, Father Kaspar Schott, did not let the manuscript lie, but, replete with references to Giordano Bruno, he republished it in Germany, out of reach of the Roman Inquisition.[11]

Since the seventeenth century, both science and religion have come a long way in recognising that healthy scepticism is an inescapable and humanising ingredient in their respective quests for truth. Without it, science and religion can tip over into extremes, where madmen of either the religious or the secular variety end up wreaking havoc on the world. Fortunately, influential thinkers are overturning the old-fashioned view that religion and science are age-old enemies. Indeed, former *New York Times* science writer Dava Sobel, whose books *Galileo's Daughter*, *Longitude* and *The Planets* have contributed enormously to the popular understanding of science, naturally refers to the Bible when she describes the origins of the universe:

> The Book of Genesis tells how the dust of the ground moulded and exalted by the breath of life became the first man. The ubiquitous dust of the early Solar System—flecks of carbon, specks of silicon, molecules of ammonia, crystals of ice— united bit by bit into 'planetestimals', which were the seeds, or first stages, of planets.[12]

---

11. Rowland, *Ecstatic Journey: Athanasius Kircher in Baroque Rome*, 199.
12. Dava Sobel, *The Planets*, (London: Harper Collins, 2005), 17.

As we leap metaphorically into the infinite cosmos, aided by satellites and gamma-ray telescopes, the expanding knowledge of the universe has not consigned faith in God to the dustbin of history but challenged it to be greater and sharper than we imagined. 'The iron in our blood was forged in supernovas,' says Aileen O'Donoghue, Professor of Physics at St Lawrence University in upstate New York. She is a Catholic, but her knowledge of the universe has not undone her belief in God, only revealed more of what that means:

> God works somewhere within the laws of physics. I look at the generosity of the universe, the richness of it, the fact that life manages to live in every environment, from the boiling pools of Yellowstone to the dry pits of Antarctica. Life is everywhere and it is so abundant. The universe doesn't tell me to believe in God, but once I believe in God, the universe tells me a lot about God—the abundance, the exuberance, the fecundity of it all.[13]

She is one of the luckier Catholic scientists, living in a world that is more accepting of a spirituality that sees God in nature; perhaps even, as the Kabbalists imagined, *in need* of nature. Like her kindred spirit, the Jesuit priest, palaeontologist and philosopher Pierre Teilhard de Chardin (1881–1955), who found God in the very layers of geological time which he unearthed, O'Donoghue's faith is confirmed in nature's other theatre, the night sky. That does not mean she can only appreciate God in so far as she can see, measure and categorise God in the cosmos. She is also at home with the sheer incomprehensibility of things like black holes, where light can be both a wave and a particle at the same time. She is not daunted by what she calls the 'weirdness' she finds, since it underscores the sheer magnitude and mystery of God. It is this acceptance of the unknown that separates her from the Creation Scientists who claim to have God's blueprint in the Scriptures, in words that are neither more nor less than their literal meaning. With the kind of faith that is truly in awe of the Creator, she suspects that the more literal-minded readers of the Bible are narrowing their vision and even blinding it to the wonders of creation.

---

13. Quoted in Renée LaReau 'Sacred Starry Night: Astronomer finds spiritual, intellectual home in the night skies', *National Catholic Reporter* (28 October 2005): 13.

Unlike Teilhard de Chardin's writings, which were suppressed by the Church during and after his lifetime, O'Donoghue's spiritual observations are available to all who read her daily reflections in *Living Faith*, a quarterly compendium of daily Scripture-based reflections. Hers is an active faith that daily struggles to practise believing, as she puts it. Like Jacob's struggle with God's messenger, the angel, it is only when faith is challenged that it grows and is alive to the world, including the infinite universe.

Yet even in this age of expanding knowledge, of which the Catholic Church has been a keen participant since Pope Leo XIII founded the Vatican Observatory in the 1890s to show that science and religion were not opposed to each other, it is not without its dissenters. While Pope John Paul II oversaw significant changes in the Church's official attitude toward science, including acknowledging in 1992 that an

Father George Coyne, Jesuit priest and former Director of the Vatican Observatory, was dismissed after twenty-eight years in his post for rebutting the theory of intelligent design.

Courtesy Father George Coyne

error had been made in condemning the seventeenth Century astronomer, Galileo Galilei in 1633, and also declaring that Darwin's theories of evolution were 'more than hypothesis,' his successor signaled a far more traditional view, perhaps, as the Jesuit priest and astronomer George Coyne, observed, returning to an earlier position. In the inaugural sermon of his pontificate, Pope Benedict XVI announced that 'We are not the accidental product, without meaning, of evolution'. When the Pope's close associate, Cardinal Schonborn, backed the teaching of intelligent design in schools, a vigorous rebuttal from George Coyne did not go unnoticed. On 19 August 2006, the Director of the Vatican Observatory retired after twenty-eight years of service.

There is an optimistic postscript to this story: the current Chief Astronomer of the Vatican Observatory, Jesuit brother, Guy Consolmagno, remembers the first words George Coyne said to him when he joined the Observatory back in 1993: 'do good science.' It seems he has, and now Guy shares a unique distinction with his former mentor. Before George Coyne died on 11 February 2020 at the age of eighty-seven they were the only two living men who had an asteroid named for them.[14]

---

14. https://www.azcentral.com/story/news/local/arizona/2020/02/16/george-coyne-priest-who-helped-establish-arizona-mount-graham-observatory-dies-at-87/4752546002/

Chapter 3
Mad About Books

'Let nothing disturb you, let nothing frighten you. All things are passing.' These words of Carmelite nun St Teresa of Avila, who died in 1582, provide much comfort to people today, along with the words of even earlier medieval mystics like Hildegard von Bingen, Meister Eckhart and St John of the Cross. Perhaps as an antidote to the materialist values and scientific verities of our present era, their writings are of growing interest today and they have been gathered together in a unique library, the only one of its kind in Australia. Thousands of volumes on religious mysticism, some in their original pressings which date back to the sixteenth and seventeenth centuries, are housed in the Carmelite Hall, built in the Gothic style of 1918, in Melbourne's Middle Park. Now renamed the Carmelite Centre, the men and women who manage its community events and provide pastoral services as well as share the library with the interested public, strikes a very different chord to the original intention of the Carmelite order, which began as a collection of hermits in the 1190s on Mount Carmel, Israel, during the religiously charged era of the Crusades.

Back then, contemplative orders and individual anchorites withdrew into their cells awaiting the Day of the Lord, certain that their prayers would ensure a swift passage to heaven. Today, contemplative orders such as the Carmelites are not in full retreat from the secular world but find ways to witness to it, not least through the broad range of books on mysticism held in the Carmelite Library. What is even more significant is that some of the books on its shelves were once censored by the Church for their heretical views, for example the writings of the thirteenth-century Dominican preacher and mystic Meister Eckhart and the early twentieth- century palaeontologist and Jesuit priest Pierre Teilhard de Chardin. There are also a number of books by Buddhists, Hindus and other non-Christians, who would have been viewed as pagans and Devil worshippers by the Church of old.

The availability of this eclectic collection is a far cry from the situation that greeted the curious monk or nun in the sixteenth century, when news of other religions, cultures and burgeoning scientific speculation stimulated a great interest in books from distant lands and a desire to collect them in private if not secret libraries. To be sure, the Vatican still issues a list of 'banned books' which are meant to be avoided by the faithful, and has even shredded volumes in which its priests and nuns have promoted ideas in stark contradiction to Church teachings. However, it is a matter for the individual cleric or lay person to decide whether to stay and be disciplined or to leave the Church. After thirty-five years as a member of the Institute of the Blessed Virgin Mary, Sister Lavinia Byrne in England elected to leave her order because her books advocating the ordination of women put her on the wrong side of the Church's teaching, resulting in the shredding of her book *Women at the Altar* (1994); and after thirty-three years, Father Paul Collins in Australia, resigned from active priestly ministry, due to his critique of the Vatican's exercise of power, which he dubbed monarchical and authoritarian in his book *Papal Power* (1997). Both cited as their reasons to leave their respective callings that they were subject to censure and 'bullying' by the Vatican's Congregation for the Doctrine of the Faith. In electing to walk away from active service in a church, they were not however leaving behind their identification as Catholics but continued with their careers as scholars and religious commentators.

In the Renaissance, such transitions were not easily achieved by the Church's independently minded religious, for whom there was no 'secular' society from which they could receive support and escape the authority of the Church. Instead, they concealed their interests and cultivated discreet friendships to avoid the Church's wrath. Indeed, Teresa of Avila's words might well have been written for the very bibliophiles who fell foul of the Church and were tortured for their curiosity and desire for knowledge.

The Medici family was among the greatest book collectors and record keepers, but it is only in recent years that its vast holdings in Florence have been organised and catalogued into anything resembling good library practice. Edward Goldberg, a former art historian teaching at Harvard University, came across the Medici collection in the course of his work and was dismayed by the disorganised state of affairs. He decided to do something about it. After unsuccessfully lobbying the Italian Government, he returned to the United States and raised the money to fund the Medici Archive Project. Since 1995, Goldberg has presided over the translation and cataloguing of three million letters, contained in 6000 volumes, covering the years 1537 to 1743. Out of it have come some remarkable studies, including his *Jews and Magic in Medici Florence: The Secret World of Benedetto Blanis* (2011) which focuses on the relationship between Giovanni de Medici and the Jewish businessman, Benedetto, who had a mutual interest in the occult, and Harvard historian Brendan Dooley's 2002 biography of a charismatic figure at the heart of the church who kept a library of his own, Orazio Morandi.[1]

In 1600, the year that Giordano Bruno met his gruesome death, the Roman monastery of Santa Prassede made an inventory of its library for papal officials, who were intent on confiscating any volumes that appeared on the Index of Forbidden Books of 1557 and later. On this occasion, no such books were found among the 200 or so in the Prassede Library. But that doesn't necessarily mean they were not there. Secreting away controversial works was, it seems, a common practice among intellectually curious monks during the period of the Counter- Reformation, when the Church of Rome was keen to reassert religious orthodoxy after the turmoil caused by the Christian reformers John Calvin and Martin Luther. The Church's efforts to stamp out the reformist influences from the German lands, as well as its curtailment of the humanistic and scientific preoccupations of the Renaissance, led to the most violent means of suppression by the Inquisition. Imprisonment, interrogation, torture and beheadings were not uncommon as the Church tried to stem the tide of heresy. Ironically enough, the clerical class was not immune to the brutality of the Inquisition. Although they were often handled with more care than was the common rabble, educated monks were most likely to come into contact with the radical ideas contained in contraband books.

---

1. Brendan Dooley, *Morandi's Last Prophecy: The End of the Renaissance* (Princeton: Princeton University Press, 2002).

Although the Prassede Monastery escaped prosecution in 1600, its fate would change with the arrival in 1613 of a new abbot from Tuscany. Up until his arrival in Rome, Orazio Morandi had enjoyed considerable freedom to pursue the study of astrology and the occult sciences with his great friend and patron Giovanni Medici. Having spent time with the powerful Medici family, the great patrons of Renaissance culture, Orazio Morandi knew only too well the necessity of cultivating influential friends. According to Brendan Dooley, he was an ambitious monk, who began as a member of the large Benedictine order, but left it for the smaller and more independent Vallombrosa order, which afforded him greater independence to pursue his interests while enjoying the protection of the Medici rulers of Tuscany. Under their influence, Morandi's appetite for 'dangerous knowledge' contained in banned books was born, and by the end of his life he had amassed a considerable private library, which was patronised by various Church officials and, indirectly, the Pope himself. And Morandi had another string to his bow, making him even more valuable in the eyes of these gentlemen: he was an astrologer.

Nothing was more intriguing to ambitious men than the power of prophecy, and no one was better suited to that mercurial art than astrologers. A mere generation before Morandi could lay claim to being the last great Roman astrologer, another prophet—perhaps the greatest soothsayer of all time—Michel de Nostredame (Nostradamus of Provence), counted among his illustrious clients the French King Henri II, the French Queen Catherine de Medici and the English Queen Elizabeth I, as well as a host of lesser royalty, dignitaries and wealthy businessmen. A biography of Nostradamus by Ian Wilson, which begins with an exploration of his French Jewish forebears, who converted to Catholicism, reveals the breadth of Nostradamus's expertise. Royal matchmaking and business advice were among his specialities, but so too was medical advice, which relied on the casting of horoscopes to diagnose ailments, prepare remedies, and to give a general prognosis of one's health.

Unlike his monastic counterparts, Nostradamus was a family man whose monthly almanacs and personal advice, the quatrains on the future of the world published as his *Prophecies*, as well as medical potions and kitchen recipes, were all a lucrative source of income. But Morandi, the abbot of Santa Prassede, could not practise his astrological arts in the open, nor could he exploit them as a source of income. Working in secret, Morandi could only hope to gain access to prized manuscripts or books from learned clients and, most importantly, secure the promise of protection from powerful friends in exchange for the astrological charts and predictions that he and his fellow monks prepared for them. Given these were officially forbidden by the Vatican in its anti-astrology Bull of 1586, Morandi needed all the protection he could get.

A cardinal rule of the astrologer who is looking for repeat business is to avoid offending the client, which rules out predicting their imminent death. When the client is Pope Urban VIII, a serious devotee of astrology and the magical arts (defying the Church's own prohibitions), the astrologer is in a particularly invidious position. Morandi was keenly aware of the challenge, since Urban VIII had already consulted one of the period's most notorious heretics, the brilliant astrologer Tommaso Campanella, who was enjoying a brief respite from prison and the unwanted attentions of the Inquisition. Campanella had been summoned by Urban VIII to advise him on how to counteract the morbid predictions made by some Roman astrologers, who had warned against the fatal astral influences of a solar eclipse followed by a lunar eclipse. (Remember that astrologers were akin to doctors, prescribing all kinds of potions and combinations of disparate items—from a virgin's hair to a crone's fingernail mixed with baby's spittle—for their clients to ward off the effects of negative astral forces.) Morandi, who was a rival of Campanella, set about preparing his analysis of the Pope's fate, but anonymously. It was, of course, greatly interesting to the Roman community, especially the anti-Urban faction, who would pay for this information. Based on the nativity chart of Urban VIII, who was born on 5 April 1568 at 1.29pm, his conclusion was inevitable: the Pope was doomed. The solar eclipse that was predicted for June 1630 would bring about the death of Urban VIII.

While other monks at Santa Prassede agreed with Morandi's prediction, there was one astrologer, a frequent visitor to the monastery, who disagreed. Raffaele Visconti, secretary to the Master of the Sacred Palace and Chief Censor for the Congregation on the Index of Forbidden Books, argued for a more positive interpretation of the chart, which mitigated the negative effects of Saturn and displayed an improvement as the year 1630 progressed. Furthermore, he noted that previous eclipses had occurred during the Pope's life, such as in 1624, with no adverse effects. According to Visconti, fatal conjunctions of the planets for Urban VIII lay far into the future. Despite this more positive outlook, Visconti's view did not prevail, and the circulation of Morandi's fatal prediction travelled far and wide— disguised, as was Visconti's alternative version, with a postmark from Lyons (Lyon). Why Lyons? According to Ian Wilson, this southern city on the Rhone River, second only to Paris in printing, publishing and bookselling, was renowned for publishing the almanacs and quatrains of Nostradamus. The Lyons postmark was undoubtedly a frequent subterfuge for those who practised the banned art of astrological prediction under the noses of the Roman Inquisitors.

Despite being unsubstantiated, it is not hard to imagine even the most sanguine ruler being alarmed by a rumour that he was fated to die during the solar eclipse—and Urban VIII was most definitely not sanguine, given the deep financial debt that he was in, thanks to the construction of St Peter's Basilica and the ongoing battle with the French Church, which had elected six rival popes in Avignon in the 14th Century. The predicted death of Urban VIII also raised the hopes of those anxious to succeed him. Indeed, one such hopeful, Cardinal Desiderio Scaglia, was delighted with the news, since his own astrologer, Francesco Lamponi—a trained lawyer and governor of the Tuscan town Saint Gemini—had made a similar astrological prediction of the pope's imminent demise. Like Visconti, Lamponi was a frequenter of the Santa Prassede library.

It would not be long before the Pope's investigators narrowed in on the monks at the Santa Prassede monastery. Perhaps, as Brenden Dooley suggests, they were steered in that direction with the help of Tommaso Campanella, who was anxious to divert suspicion from himself. After some initial use of torture, the monks of Santa Prassede were offered immunity in exchange for information about the goings-on in the monastery. They soon spilled the beans and secret caches of banned books were uncovered, including works by Machiavelli and Copernicus, as well as erotic writings and esoteric magical texts by Jewish Kabbalists. Eventually, there was enough incriminating evidence to arrest the abbot who oversaw it all. Orazio Morandi was imprisoned and subjected to a trial. But, perhaps inevitably, the trial of Morandi, an eloquent and well- connected fellow, would lead to embarrassing revelations about Roman high society, including key figures in the Church. Lending records revealed that a cross-section of distinguished churchmen—including three cardinals, the famed artist Bernini, and a host of noblemen—enjoyed the library's holdings.

Fortunately for those implicated, all such concerns were laid to rest when Morandi died suddenly on 7 November 1630 in Tor di Nona prison in Rome. It was widely believed that he was poisoned, since the official reason for his death was noted as a malignant fever of unspecified origin. Either way, it would not have been difficult to cause a 'natural death' in such primitive conditions, where lack of sanitation might easily lead to food poisoning or turn a festering wound septic. What Morandi undoubtedly lost sight of in his breathtaking entree into high society was that he was expendable in the overall scheme of things. Although he was saved from the ghastly torture of death by burning, his books were not. Cast into the fires of the Inquisition, one of the great libraries of the Renaissance would never reveal its secrets to posterity.

Across the English Channel, and about a generation before Morandi was born, another private library was in the making. Its keeper had remarkably similar interests to Morandi, but he did not have to contend with the Papal police. John Dee, who has been variously described as Queen Elizabeth's private astrologer and Britain's foremost magus, was a man of precocious intellect who had a studied ability to remain religiously ambivalent, not to say ambiguous. This was a very handy attribute that few courtiers managed to attain in a period of English history that saw many loyal subjects and royal confidants entangled in the internecine religious conflict between Catholics and Protestants, only to end up in the Tower of London awaiting execution. Indeed, John Dee's own father, Roland, a prosperous Protestant businessman and tailor to Henry VIII, was imprisoned for a time during Queen Mary's campaign to rid England of Protestant heretics in the 1550s, and his ruination had undoubtedly pressed upon his son the need to navigate religious politics extremely cautiously.

John Dee's real interests were far more esoteric, and included just about every form of knowledge from science to spirits. That Dee managed to escape the hangman's noose is amazing given how close he sailed to danger. If it wasn't the shady company he kept, it was his active involvement in magical divination, prohibited by both state and divine law, which could easily have put him away forever. Yet

Dᵉ Dee avoucheth his Stone is brought by Angelicall Ministry.

John Dee was one of the most fascinating and formidable men of Elizabethan England. His huge library and encyclopedic knowledge gave him considerable influence over the Queen. His reputation as a necromancer and his involvement with alchemy, however, put him at dangerous odds with both state and divine law.

for all that, Dee was a highly informed man, part scholar (he was a graduate of St John's College, Cambridge, and a founding Fellow of Trinity College in 1546) and part autodidact, whose formidable knowledge of geography, astronomy and history made him a valued adviser of Queen Elizabeth. From astrological predictions of a personal nature to the political and economic analysis of Britain, Dee's advice to Elizabeth reflected the diverse interests of his sovereign and drew on his vast knowledge of the books in his collection. According to his biographer, Benjamin Woolley, John Dee's library was one of the largest in Europe, containing books on every topic, including demonology, dreams, horticulture, earthquakes, Islam, pharmacology, veterinary science, astrology and his special passion, mathematics.[2] In fact, bulky volumes and hundreds of manuscripts

---

2. Benjamin Woolley, *The Queen's Conjuror: The Science and Magic of Dr John Dee* (London: Harper Collins, 2001).

filled his mother's cottage, which he inherited. To accommodate his growing collection, he expanded the house several times and transformed it into a library with a public reading and copying room, as well as a laboratory, where he conducted alchemical experiments in search of the holy grail of the occult world, the recipe for gold. Like everyone else who attempted it, he was unsuccessful.

John Dee was neither a professing Catholic nor a fully committed Protestant,[3] but he searched for God's truth in the vast ocean of books he collected. He was acutely aware, however, that, next to disloyal courtiers, books were the most endangered species in the religious tug of war between the Catholic and the Protestant contenders for the English throne. On the Continent, with the Inquisition in full swing, dangerous books went up in smoke just as heretics were burned at the stake. Even in England, before Elizabeth I was on the throne, the Catholic Queen Mary I ordered Archbishop Thomas Cranmer to burn at the stake for his reforming tendencies, which included the translation of the Latin Bible into English. Under Mary's reign, both the English Bible and Archbishop Cranmer's Book of Common Prayer would have been hard to come by, and possession of them would have been the swiftest way to death. Yet, as a passionate bibliophile, Dee embarked on a project which, one way or another, would occupy most of his life: to collect works for a national archive that would encompass all the knowledge of the world, from antiquity to his own day. Although he never received official support for his project, either from Queen Mary or her sister Elizabeth, he travelled to Europe several times on buying expeditions and funded his expensive and voracious habit by selling his services as an astrologer and teacher, interpreter of dreams and even a medical consultant.[4]

It appears Dee went to unethical and even criminal lengths to acquire precious volumes.[5] But it was *Steganographia* by the German abbot Johannes Trithemius which John Dee risked all to acquire. Trithemius, who was a student of the Cabala (the Christianised version of the Jewish Kabbalah), was the founder of modern cryptography, the science of codes, and also a believer in magic, which he distinguished from mere superstition, the stuff of witches and wizards, whom he

---

3. Woolley, *The Queen's Conjuror: The Science and Magic of Dr John Dee*, 45.
4. Wooley's book documents all these aspects of Dee's life, which made him a very valuable courtier.
5. Woolley, *The Queen's Conjuror: The Science and Magic of Dr John Dee*, 48.

castigated. These subtleties were lost on his critics, however, who accused him of 'trafficking' with demons, in the same way that Dee's own magical explorations put him under similar suspicion. In short the Benedictine scholar had all the qualities that intrigued John Dee, who already owned his first work on cryptography, *Polygraphia*. Dee immediately travelled to Antwerp on discovering that there he would find a copy of the *Steganographia*. He hoped that it would provide the key to deciphering other mysterious texts believed to contain the language spoken by Adam, the first man. It was not this esoteric value of Trithemius' manuscript that he promoted to his patrons, however. The court of Elizabeth I was greatly interested in the science of cryptography for one reason alone: as an aid to developing its fledgling spy network in Europe.

While John Dee made his knowledge available to the royal court, it was science and magic that interested him above all, and these seemingly contradictory fields were not that far apart in his day. For instance, he had a theory that everything in the universe emanated 'rays' which could be mathematically measured and might thereby reveal the true distances between celestial bodies. Basing his theory partly on the forces of attraction and repulsion in magnetised ore, Dee's ideas—set out in his 1558 book of astrology, *Propaedeumata Aphoristica*—anticipated Newton's theory of gravity by a hundred years. But his vast interest in what we would call natural science or observable natural processes, including the ebb and flow of tides, was curiously merged or inseparable from a profound belief in magic; that is, the harnessing of unseen forces. Dee believed that these forces were, in fact, God's divine creation and, as such, they were natural yet beyond simple rational understanding. Today we might recognise similarities in Benjamin Franklin's discovery of electricity and his ability to conduct the current in a light bulb. That discovery would be three hundred years in the future, because for Dee the unseen forces were more elusive and seemed to exist in an in-between-world that he yearned to enter himself. He believed that Magic was the key.

This is not to suggest that Dee was oblivious to practical concerns. He was often in need of money, and in one instance he hoped to secure a royal pension in exchange for the deployment of his divination expertise to secure buried treasure, specifically gold and other valuable metals, for the royal coffers of Elizabeth I. He failed to secure that commission, but in other ways he would prove to be a useful guide to Her Majesty. Dee's interest in geography and history, combined with his considerable knowledge of the navigational sciences (extending even to the size and make-up of China's considerable naval fleet), came together in several works in which he argued for a 'Brytish Impire' that would extend to foreign lands and overturn Pope Alexander VI's proposed division of the New World between Portugal and Spain. This was undoubtedly music to Queen Elizabeth's ears, and Dee outlined his case in *Brytanici Imperii Limites*, which he wrote for her. Dee, who was a friend of Sir Walter Raleigh, appears to have had some influence over Elizabeth, who was encouraged to call herself Empress as well as Queen.

The uses to which John Dee put his encylopaedic knowledge were of great help in securing him a real friendship with Elizabeth I. She not only turned to him for advice and even visited his home in Mortlake, but also allowed him considerable freedom to pursue his more arcane interests. When, for example, he gave the queen his negative prognosis of her proposed marriage to the French duke of Anjou, she would have been impressed, at the very least, that one year later the unlucky gentleman had died as Dee had predicted.

The reputation of John Dee as a necromancer, able to speak to angels and spirits and to see into the future, was known on the Continent, and one of the people who sought him out was a pretender to the Polish throne, Lord Albert Laski, himself an enthusiastic dabbler in the occult. On his behalf, Dee contacted 'the angels' for information about Laski's prospects, and apparently received an endorsement from the angel Raphael. But Dee was also in contact with other spirits, including a happy young girl named Midimi and a morose fellow named Murifri. What ensued in this extraordinary period in Dee's life, which is detailed in his logbook, was a desperate attempt to confabulate his predictions of Elizabeth's future empire and Laski's hopes for the Polish kingdom into a vision that saw the two of them working together to secure Europe for the British Empire. Elizabeth herself was interested in Laski, and it was perhaps with her approval that Dee embarked on a trip to Poland and Bohemia in 1583 to investigate the situation in person.

Having commenced his journey from London in the dead of night, John Dee's travels to Central Europe have all the intrigue of espionage in lands that were seething with Catholic/Protestant rivalry against the backdrop of the even greater threat of the Ottomon

Empire's incursions into Europe. But other, more lucrative interests beckoned, including an invitation from the immensely wealthy Bohemian nobleman Vilem Rozmberk, who shared Dee's interest in alchemy. Rozmberk paid Dee and his assistant, Edward Kelley, to undertake experiments in his personal laboratory for the purpose of, among other things, finding a cure for his infertility. The two Englishmen took up residence with their wives in Rozmberk's castle in the Bohemian town of Česky Krumlov, and proceeded to develop a program of mediumistic 'actions', dutifully noting down outlandish visions and messages. Eventually the ambitious Kelley, who was on the run from the law yet much in demand for his services as a necromancer, absconded to Prague. After six years on the Continent, Dee left for England.

When the weary Dee returned to his home in 1589, he was in for a shock. His library, his laboratory and his significant collection of scientific instruments, including clocks, telescopes and a sea compass as well as his crystal balls and magic mirrors, had been plundered. His laboratory vessels and concoctions had also been broken and looted. Above all, his painstakingly acquired library was severely depleted, with Dee estimating that 500 volumes had been taken or that his brother in law, Nicholas Fromond, with whom he entrusted the care of his library had 'unduly sold it presently upon my departure or caused it to be carried away'.[6] Whether it was a peasant mob that ransacked his home or some clever thieves who offloaded their precious cargo to receivers was never ascertained but many of Dee's books would resurface with his name removed and replaced by others, especially one Nicholas Saunder, a frequenter of Dee's circle of scholars. A catalogue of John Dee's library compiled by Julian Roberts and Andrew Watson suggests that the missing books represented only a small portion of the library that was catalogued in 1583 and considered to be England's largest and, on scientific subjects, its most valuable. No less than The Royal College of Physicians now owns over a hundred of Dee's volumes, the largest collection of books stolen

6. The Lost Library of John Dee on display at the Royal College of Physicians exhibition 18 January 2016–28 July 2016. https://www.rcplondon.ac.uk/news/lost-library-john-dee#:~:text=The%20Royal%20College%20of%20Physicians, Dee's%20books%20in%20the%20world.&text=He%20claimed%20to%20 own%20over, London%2C%20on%20the%20River%20Thames.

from his library, and in 2016 invited the public to inspect them in an exhibition that lasted six months.[7]

John Dee lived a long and productive life into his 80s, but he would not escape falling victim to scandalous rumours that were aimed at discrediting him in the eyes of God-fearing Christians. He was accused of sorcery—which, of course, was how most people would have described his work, although he genuinely believed he was resorting to rational knowledge of a scientific nature in order to communicate with spirits and angels. Indeed, as a member of Sir Walter Raleigh's 'secret ungodly academy', Dee counted himself among those who sought 'the vital warmth of freezing science', according to a prominent member and co-founder of the School of Night, George Chapman. His poem *The Shadow of Night*, published in 1594, is an occult encomium to the Goddess of Night and is full of Cabbalistic allusions which he hoped to advance. Unlike the conditions that prevailed in Rome, where the Inquisition quashed un-Christian activity with ruthless self-interest, England afforded Dee, the court adviser, astrologer and alchemist, the opportunity to go about his magical experiments unmolested. Perhaps he was just fortunate to be in the powerful Queen's favour—nevertheless, he fully expected to die in his bed, which he did in 1609.

His library did not reveal all its secrets until a little over thirty years later, when a private chest containing his secret magical writings surfaced, and its contents were made available to lawyer, astrologer and antiquarian Elias Ashmole. An enthusiastic alchemist himself, Ashmole pored over the works, but they were far from easy to understand. To some they appeared to be the ravings of a madman, and Dee's reputation suffered. Even today, Dee is regarded as a man who was simultaneously brilliant and 'away with the faeries'. The truth was that as a Renaissance man, schooled in a wide range of subjects, from the sciences and philosophy to astronomy and the occult mysteries, he was not exclusively committed to one against the other, but believed that his lifelong researches would show that they were all somehow related, even enmeshed, rather than diametric opposites. That he would never have to demonstrate clearly, either philosophically or scientifically, how they were all related, resulted in a dizzying confusion of ideas and opaque concepts. Chapman's poem,

---

7. Ibid. 4 Sept 2020

The Shadow of Night, for example, still confounds the interpreter.[8] But to Dee, as to Morandi, it was not God and the ethereal world of spirits that stood in the way of every kind of new knowledge, but the Church. On account of its claim to know the final truth and its exercise of absolute authority, the Church was bound to act harshly to protect itself from the unbridled inquiries of freethinkers like Morandi and Dee.

---

8.   http://www.esoteric.msu.edu/VolumeVI/Chapman.htm 4 Sept 2020

Chapter 4
The Ten Lost Tribes

The Indians are of a browne colour, and without beards, but in
the new world, white, and bearded men were found, who had
never commerce with the Spaniards; and whom you cannot
affirme to be any other than Israelites.[1]

His portrait was both etched and painted by Rembrandt. He was
a distinguished scholar, a successful publisher, and was a rabbi at
Neveh Shalom, the new synagogue in Amsterdam. Menasseh ben
Israel (1604–57) was also a proponent of the view that the newly
discovered American Indians were none other than the lost tribes
of Israel. Fluent in several languages, including English, he made his
views known in a book, *The Hope of Israel* (1650), which he dedicated
to 'the High Court, the Parliament of England and to the councell of
State' and sent to Oliver Cromwell, the founder of the fledgling but
short-lived English republic. By all accounts it was well received by
the Protestant leader, who knew his Bible and who, along with many
other Christians and Jews, had speculated on the fate of the ten tribes
of Israel, which had been vanquished and exiled by the Assyrians
several hundred years before Christ.

There were twelve tribes of the people Israel, of which ten lived in
the northern kingdom of Israel and two in the southern kingdom of
Judah. The Bible tells the story of the northern kingdom's conquest
in 734 BC by the king of Assyria, Tiglath- Pileser III, and twenty
years later by Shalmaneser V, who exiled the ten tribes to Assyria,
where, as the writer of 2 Kings records in 17:6, 23, they are 'until
this day'. Tablets inscribed in the Akkadian language also report the

---

1. Menasseh ben Israel, *The Hope of Israel*, Sect 7:23.

victory over Samaria, the capital city of Israel, specifying that 27,900 people were 'carried away'. Although a similar conquest and exile of Judah occurred fifty years later at the hands of the Babylonians, they eventually allowed the exiles to return to the land of their forefathers in the sixth and fifth centuries BC to rebuild the temple at Jerusalem and regain sovereignty. The ten tribes of Israel, however, remained something of a mystery, as there is no record of their return. Were they assimilated into the host culture or did they remain a strong community beyond the Euphrates River, as the ancient Jewish historian Flavius Josephus believed?

Certainly the apocryphal book of the Bible, the Apocalypse of Ezra (2 Esdras), which was probably composed in the late first century AD, held sway over many who pondered the question. In 13:40–45 it tells of the ten tribes taking counsel among themselves and venturing 'into a further country, where never mankind dwelt, that there at least they might keep their statutes . . . [beyond] the Euphrates . . . [in a land] named Arzareth'.

A portrait by Rembrandt of Menasseh ben Israel, a rabbi at the Neveh Shalom synagogue in Amsterdam, who believed that the newly discovered American Indians were the lost tribes of Israel.

For Menasseh ben Israel the 'discovery' of the ten lost tribes in the Americas had more than academic significance. During the era of emerging Protestantism, when the Reformed Christians were keenly interested in reading the Bible for themselves (unlike Catholics who were discouraged from doing so), there arose a strong interest in the fulfilment of prophecies which would bring about the millennium, the 1,000-year reign of the returned Christ as described in the Book of Revelation, the last book of the Christian Bible. Central to Christian understanding of the prophecies was the dispersal of the people Israel prior to their restoration, as foretold in the Book of Daniel. Menasseh had an ulterior motive in sending his book to Oliver Cromwell, whom he also met in person to plead his case. No Jews had been allowed to live in England since the year 1290, when they were all expelled under King Edward I, who had previously imprisoned and ransomed about 3000 of them in 1287. Menasseh, who had many Christian friends and whose own father was forced to convert to Christianity in his native Portugal, hoped to convince the new English ruler that allowing Jews to return to England would contribute to the dispersal of Jews around the globe, and hence fulfil the conditions necessary for the Christian millennium.

Menasseh ben Israel was certainly not the first to make a connection between the native peoples in the New World and the ancient Hebrews. The Spanish had already done so for quite different reasons to those adduced by the rabbi; they were keen to understand how it was that the alien people they encountered could have developed such advanced practices of sacrifice and ritual purity, and beliefs in a creator god of the heavens and the earth. Surely the God who created the whole world, as written in the Bible, had not overlooked them— or, worse, abandoned them to Satan so that he might possess half the world for his diabolical purposes! On the contrary, they must be a remnant of the ancient Hebrews.

In 1492 Christopher Columbus sailed to the New World, taking with him a Hebrew-speaking Jewish convert to Christianity, Luis de Torres, in the hope that Torres would be able to converse in Hebrew with the inhabitants. It should be remembered that Columbus thought he was sailing to China, and on the island of Cuba, which he believed to be the mainland of China, he expected to meet the great Kublai Khan. As it turned out, he was nowhere to be found. But why would Kublai Khan speak Hebrew in the first place? The answer leads back to the Apocalypse of Ezra—namely, that the ten lost tribes of Israel travelled beyond the Euphrates to become a great civilisation in the east.

One of the most infamous Spaniards, the Inquisitor General, Tomas de Torquemada (1420–98), also turned his attention to the inhabitants of the New World, but, as one would expect, he saw the Devil everywhere. What seemed to be Jewish and Christian rites practised by the natives, this 'hammer of heretics' took to be the Devil up to his old tricks, setting up counterfeit practices 'in order to preserve a pretended and imaginary glory among those poor blind and deluded Indians'.[2] Torquemada's treatise relied on another authority, Las Casas, for a list of the similarities between Indians and Jews. Las Casas stated, for example, that the American languages spoken in Jamaica, Cuba and elsewhere were nothing but dialects and corruptions of Hebrew and Yiddish. 'Cuba', he believed, was perfect Hebrew for the 'helmet' which Saul gave to David, and thus the name was proof that the island was discovered by Jewish chieftains, who named it after the helmet worn by the first ruler of the island. Convoluted reasoning, to be sure, but in this fashion Las Casas identified many place names and foods with Jewish origins.

For Torquemada and other missionary-minded Christians, the alleged Jewish origins and practices of the Indians only confirmed that they were in league with the Devil and were in need of the Gospel. Some even held the view that Saint Thomas the apostle, who is said to have travelled to India, went further afield and reached America, but his message was rejected by the Indians just as the Jews had rejected Jesus. Thus did the Indians and the Jews reveal their essential unity! Even the barbarous practices of the native Americans, such as cannibalism and scalping, were explained with reference to passages in the Bible. For instance, in a series of verses beginning 'if you walk contrary to me . . .' the Book of Leviticus spells out the consequences for the Israelites of failing to adhere to the discipline and statutes laid down by God, who says in 26:27–29, 'And if in spite of this you will not hearken to me . . . then I will walk contrary to you in fury, and chastise you myself sevenfold for your sins. *You shall eat of the flesh of*

---

2. Torquemada, *Monarchia Indiana*, Madrid, 1723, quoted in Lynn Glaser, *Indians or Jews? An introduction to a reprint of Manasseh ben Israel's* The Hope of Israel, Roy V (California: Boswell Publisher, Gilroy, California, 1973), 20.

your sons and you shall eat the flesh of your daughters [my emphasis].'
This gruesome threat was interpreted by Torquemada and others as
proof that the ancient Hebrews had resorted to cannibalism and that
the Indians were none other than the backsliding Jews of old.

There were also proponents among the English of the thesis that
native Americans were Jews who settled America. Roger Williams,
who established Rhode Island, Cotton Mather, the influential Puritan
pastor of Boston, and Jonathan Edwards, the great evangelical
preacher, all espoused the theory. William Penn (1644–1718), the
Quaker who founded the province (and later state) of Pennsylvania,
was convinced that it was possible for the ten tribes to have travelled
from Asia to America, as God intended to make 'the passage not
uneasy for them'. Furthermore, there was the proof of his own eyes:

> For their origin, I am ready to believe them of the Jewish race.
> I mean. Of the stock of the Ten Tribes . . . I find them of like
> countenance, and their children of so lively a resemblance,
> that a man would think himself in Duke's-place or Bury-street
> in London, when he seeth them. But this is not all; they agree
> in rites, they reckon by moons, they offer their first fruits, they
> have a kind of feast of tabernacles; they are said to lay their
> altar upon twelve stones; their mourning a year, customs of
> women, with many things that do not now occur.[3]

There were other traits Penn claimed to recognise, such as the
distinctive nose of both Jews and Indians, and their deep- throated
manner of speaking (a reference to the guttural sounds of Hebrew).
These 'similarities' became a standard part of the story that lasted for
almost 350 years, and the occasional critical appraisals and humorous
satires did little to dent the popular belief, which was reproduced
in dozens of histories of the Americas, including South America,
Mexico and North America. One contributor to this enthusiasm was
the colourful figure Major Mordecai Manuel Noah (1785–1851),
diplomat, journalist, politician and advocate of an American Jewish
utopia that would incorporate the American Indians. In September
1825, Noah dedicated Ararat, downstream from Buffalo in upstate
New York as 'A City of Refuge for the Jews'. It was a great public

---

3. *William Penn, His Own Account of the Lenni Lenape or Delaware Indians*, 1683,
    by Albert Cook Myers (Philadelphia, 1737) in Glaser, *Indians or Jews?*, 46.

ceremony with music, processions and a speech by Noah, in which he proclaimed the future mingling of Indians and Jews: 'If the tribes could be brought together [they] could be made sensible of their origin, could be civilised, and restored to their long lost brethren, what joy to our people!'

Few of the Jews either in America or in Europe were enthusiastic about Noah's plan,[4] but his idea would find currency amongst another group. His address was reprinted in the *Wayne Sentinel*, the hometown paper of Joseph Smith, on 11 October 1825, and influenced the man who would go on to found the Mormon Church. Most of the enthusiasts, however, were like Elias Boudinot, motivated by millennarian beliefs that Christ's reign was at hand, which necessitated an urgent campaign to Christianise the natives. Boudinot, who was a friend of President Thomas Jefferson and director of the United States Mint, authored *Star in the West* in 1816:

> There is a possibility that these unhappy children of misfortune may yet be proved to be the long lost tribes of Israel. And if so, that though cast off for their heinous transgressions, they have not been altogether forsaken; and will hereafter appear to have been in all their dispersions and wanderings, the subjects of God's divine protection and gracious care.[5]

Boudinot relied on other similar accounts of the Indians as Jews, and in turn he was a source for many subsequent books on the same topic. It was a genre of history writing which may seem improbable today, and even insulting when one reads that Indians were referred to as 'unhappy children of misfortune' (much of the unhappiness occurring at the hands of the Europeans, one could add), but it is also thoroughly understandable from a religious and even a secular point of view. The world really was a mysterious place, with peoples, languages and customs never before encountered—yet how to explain their alien appearance if they were all 'God's children'? More importantly, how was one to behave towards them: with compassion

---

4. Mordechai Noah would continue to champion the 'Indians are Jews' thesis long after Ararat failed to be established, and his 1837 publication, *Discourse on the Evidence of the Emerican Indians Being the Descendants of the Lost Tribes of Israel*, New York, 1837, contributed to the general interest in the idea.

5. Elias Boudinot, *Star in the West* (New Jersey: Trenton, 1816), quoted in Glaser, *Indians or Jews?*, 54.

or fear? It is probable that both sentiments motivated the attempts to clarify the origins of the native Americans as the ten lost tribes of Israel, since it was a way of explaining their fearsomeness while simultaneously drawing closer to them by building a bridge to a common past. There are contemporary parallels in the way people have imagined aliens from outer space as being initially fearsome and subhuman creatures, but later discovering they have human feelings and even high ideals. This theme is reiterated in a host of Hollywood classics such as *The Day The Earth Stood Still* (1951), *Close Encounters of The Third Kind* (1977) and *E.T. The Extra-Terrestrial* (1982).

The legend of the ten lost tribes in the many published accounts was not just an exercise in religious thinking and myth making, but a genuine attempt to explain the past using a kind of primitive form of anthropology before the discipline existed, with the help of the equally undeveloped science of language or philology. Beginning with the original account of the ten tribes exiled to Assyria, which all Biblical scholars today agree occurred, they took up the story where the Bible left off and conjectured a migration eastward, across the ocean to the Americas. With raw enthusiasm, they genuinely believed that they had discovered a people forgotten by time, yet amply explicable by the sources at their disposal, including the eye-witness accounts of intrepid travelers and explorers who believed they had met 'Hebrew-speaking' inhabitants in remote South American villages. Indeed, one such account by Antoine Montezino, who claimed to have met a community of Jews in Peru that recited the most sacred Hebrew prayer, the Sh'ma ('*Sh'ma Yisroel, Adonai Alohenu, Adonai Ekhad*', Hear, oh Israel, the Lord is our God, the Lord is One), was a great influence on Manasseh ben Israel, who as we have seen, was one of many Europeans who had a reason to believe that the Indians were Jews.

On occasion these days the shoe is on the other foot, so to speak. Rather than Europeans trying to claim the native Americans as Jews, there are cases of indigenous peoples from different continents and countries—including Africa, Japan, India Polynesia and New Zealand—who claim that they are descended from the lost tribes of Israel, as I shall discuss later in the chapter. In most cases, historians have shown that it is a belief which has entered their consciousness by way of Christian missionaries and has little basis in historical fact. DNA technology is providing conclusive evidence, as Dr

Simon Southerton, a molecular biologist with the CSIRO, recently discovered. In *Losing a Lost Tribe*, Southerton, who was raised as a Mormon in the Church of Jesus Christ of Latter Day Saints and also held the position of bishop in the Mormon Church, rebutted the teaching of the Book of Mormon, which claims the native Americans and Polynesians are descended from Israelite tribes who migrated to the Americas.[6] Southerton referred to a study in which more than 7000 native Americans were DNA-tested; it was proved that 99 per cent of them came from Asia. Although he is quoted in a newspaper article[7] as having not been an active member of the Church for seven years, he was nonetheless summoned before a Church disciplinary hearing. In the preface to his book, however, Southerton indicates that there is some confusion in the Mormon ranks:

> Most LDS scholars today want to limit the Israel colonization of the region to Mesoamerica, while a growing subset shrinks the Book [of Mormon's] claims even further. But seemingly oblivious to this revisionist scholarship, LDS leaders continue to teach that all or most Native Americans and Polynesians are literal descendants of the Israelites described in the Book of Mormon.[8]

The Mormons are a well-organised and wealthy Church and, like many other Churches, they engage in missionary activity among indigenous peoples around the world. But that is where the similarities to most Christian denominations end, for the Mormons believe that the Promised Land is America. The legend of the ten lost tribes finding their way to the New World had more than three hundred years of currency by the time Joseph Smith came on the scene. In 1827, Smith wrote the Book of Mormon, which is one of the holy scriptures of the Church he founded. Smith's interest in the legend had recently gained support from new 'physical evidence' of the ancient Hebrews' presence in the land. Newspapers at the time, including Smith's local *Palmyra Herald*, reported on the discovery of 'mounds' of mysterious origin, which were attributed to a sophisticated people

---

6. Simon Southerton, *Losing a Lost Tribe: Native Americans, DNA and the Mormon Church* (Salt Lake City, Utah: Signature Books, 2004).
7. *Sydney Morning Herald*, 21 July 2005.
8. Southerton, *Losing a Lost Tribe: Native Americans*, xvi.

that had inhabited America prior to the Indians. These giant heaps of skeletons were of great interest to, among others, the governor of New York and amateur historian De Witt Clinton, who estimated that the three mounds he examined in Canandaigua, New York, were a thousand years old.[9]

William Henry Harrison (1773–1841), who would become the ninth president of the United States, was one of many who speculated on the identity of the mound builders and what had happened to them. His theory, which became a leading argument in the field, was that the mounds were the last monuments of an ancestral race which had been mercilessly vanquished by the American Indians. But Harrison, whose nickname was 'Tippecanoe', was not exactly an unbiased observer, given that he first rose to fame as a war hero when he defeated the American Indians at the Battle of Tippecanoe in 1811. The mystery of the mounds attracted many commentators from all walks of life, from prominent men like Harrison to virtual unknowns like Joseph Smith, but they had a common motive: many Americans were keen to tell the story of their nation and establish an ancient Biblical lineage for themselves, especially after they had broken ties with Mother England in the final and decisive war of 1812, in which Harrison served as a general.

Like Harrison, Joseph Smith was keenly interested in explaining the origin of the mounds and writing a history of Ancient America, which, according to his mother, he imaginatively recounted to his family in 'amusing recitals'.[10] Soon he would share his stories with a wider public which was eager to find the true origins of the magnificent land they called home and had still not fully settled. Like Europeans, who were plundering artefacts, manuscripts and engraved tablets from ancient ruins and secret caves in the Middle and Far East in the hope of shedding more light on their unique place in history, so Americans were also expecting to find tangible evidence and sacred texts that would reveal the antiquity of the so-called New World and their divine purpose in it.

After reading accounts in his local newspapers of a full history of the Indians that was found in a hollow tree in Canada and of brass

---

9. Glaser, *Indians or Jews?*, 64.
10. Fawn Brodie, *No Man Knows My History* (New York: Vintage, 1948), 35, citing Lucy Mack Smith, *Biographical Sketches of Joseph Smith the Prophet and his Progenitors for Many Generations*. Liverpool, England, 1835, 85.

plates discovered buried along with other items in upstate New York, Joseph Smith went one better. In 1827 he claimed to have found the Biblical history of the American people inscribed on a set of golden tablets, written by Mormon and buried by Moroni in 438 AD under a hill in an old Indian battleground in Comorah, New York. And as one would expect, Smith claimed they were written in the 'Reformed Egyptian' language. Why Egyptian, of all tongues? Because the British had only recently acquired the famed Rosetta Stone, on which was inscribed a decree by Egyptian priests that was written in three scripts. The large stele caused a sensation because its threefold inscription was the key to deciphering the Egyptian hieroglyphs. The Rosetta Stone, which was discovered by the French in 1799 in the town of el-Rashid (Rosetta), was one of the prizes that the French handed over to the British, along with other antiquities, as part of the provision of the Treaty of Alexandria (after a battle the British won) in 1801.

Joseph Smith claimed to have found proof that two families of the lost tribes of Israel had made their way to America. On this basis he founded the Church of Jesus Christ of Latter Day Saints—or Mormons—which has ten million adherents in the United States today.

Joseph Smith's Book of Mormon, which he published in 1830, tells the story of two Hebrew families that were the founding fathers of the American people. Nephi, the son of Lehi, left Jerusalem in 600 BC and, with his father, brothers and a few others, sailed to America. His older brothers, Leman and Lemuel, were cursed by God with red skin for their unremitting sinfulness, and their descendants became the Indians. Nephi's progeny, on the other hand, were white and peaceful. Nevertheless, the two peoples—red and white—fought each other for a thousand years, piling up mounds of the dead after each battle. This was Joseph Smith's explanation of the mysterious Indian mounds. As for finding the tablets in the first place and then knowing how to read them, that too was easily explained: an angel had given the golden tablets to him, and he 'translated them by the gift and the power of God'. However, many historians have pointed out that if Smith's story came to him by way of an angel, then the angel must have read the book by Ethan Smith (1762–1849), *View of the Hebrews; or the Tribes of Israel in America*, published in 1823, which has remarkably strong parallels to the Book of Mormon. Joseph Smith subsequently appended another 'history' onto his original one (having found additional tablets), which picked up on a popular theory that proposed some people had migrated to America much earlier than the time of the ten lost tribes, having escaped the destruction of the tower of Babel as related in the Book of Genesis. In around 2500 BC, these people, whom Smith called the Jaredites, sailed to America, like Noah on his Ark, bringing with them the entire flora and fauna that flourished there.

There was much about Mormonism that caused consternation among Christians, not least that Joseph Smith wanted to establish a Mormon theocracy in America and that he continued to receive revelations introducing new beliefs, such as in 1843 when he was given divine permission to practise polygamy. Although Smith never openly cohabited with his thirty or so wives, his successor, Brigham Young, was more open about plural marriage and had fifty-six wives and as many children by nineteen of his connubial wives. Christians accused Smith not only of usurping the Bible but substantially rewriting it; in 1836 he claimed that 'Moses appeared before us and committed unto us the keys of the gathering of Israel from the four parts of the earth, and the leading of the Ten Tribes from the land of the north' (referring to their being hidden somewhere in the

Arctic).[11] Despite the long, drawn-out conflicts with the Mormons, which resulted in the imprisonment and lynching of Joseph Smith, there are ten million adherents in the United States today, and they do much to keep alive the belief that America is the home of the ten lost tribes of Israel. Among its Articles of Faith is the belief 'in the literal gathering of Israel and in the restoration of the Ten Tribes' and 'that Zion [the New Jerusalem] will be built upon the American continent'.

The Church of Jesus Christ of Latter Day Saints was forced to relinquish its practice of polygamy in 1890, on threat of losing its property to the United States Government, but news stories periodically reveal that some breakaway groups, comprising about 30,000 people in America, never gave it up. The TV series Big Love (2006–2013) was inspired by the Fundamentalist Church of Jesus Christ of Latter Day Saints, for example, which still practises polygamy, and occasionally finds itself in trouble with the law.[12]

The wealth and establishment of the Church (Brigham Young University in Provo, Utah, boasts the highest enrolment of any American university) and its almost total identification with the state of Utah have produced a self-conscious aspiration to be a mainstream Church on par with other denominations. Sociologists of religion have largely agreed, and now rarely classify Mormonism as a sect or cult, arguing that it has outgrown its controversial origins and authoritarian characteristics, while ex-members and critics say the contrary. Over the years the Mormon Church has diluted or reinterpreted some of its more objectionable beliefs, such as the racist notions which held that Black Americans and the American Indians were sinful and cursed races, who descended from Cain and sided with Satan in a heavenly war with Christ, and that this was the cause of their disadvantages, which included being barred from priesthood in the Church. Today a small number of African Americans are Mormons, although in Africa, where the Church is growing, black Africans are dominant in Mormon churches.

---

11. 'Some Members of Polygamy Sect Fleeing as Law Closes In', in *USA Today* (12 April 2006).
12. The Doctrine and Covenants of the Church of Jesus Christ of Latter Day Saints, Section 110:11. The Official Scriptures of the Church of Jesus Christ of Latter Day Saints, 2006. Intellectual Reserve Inc. Website owned and operated by the Corporation of the President of the Church of Jesus Christ of Latter Day Saints.

There is a reason why one man is born black and with other disadvantages, while another is born white with great advantage.

The reason is that we once had an estate before we came here, and were obedient, more or less, to the laws that were given us there.

Those who were faithful in all things there received greater blessings here, and those who were not faithful received less ... There were no neutrals in the war in heaven. All took sides either with Christ or with Satan. Every man had his agency there, and men receive rewards here based upon their actions there, just as they will receive rewards hereafter for deeds done in the body.

The Negro, evidently, is receiving the reward he merits.[13]

While the Mormon Church today disavows its earlier racist doctrine, there are a number of vaguely Christian groups which are founded on the belief that they are descended from the ten lost tribes of Israel in order to actively promote a virulent racist ideology. The Christian Identity Movement is one such collection of groups, including the Ku Klux Klan, the Church of the Aryan Nations, The Order, and a host of other independent 'churches' that have become newsworthy for their violent racism and lurid hate literature. Common to many of these groups is the belief that in the Garden of Eden, Eve had sex with Adam and with the Devil, resulting in two lines of people, one being the Israelites and the other being people of the Devil. The Israelites migrated to Britain and then to America, and are today the white Protestants of the English-speaking world, while the Devil or Serpent people include the Jews as well as Asians and other non- white people. According to Identity beliefs, Cain was born from the Devil or Serpent line, and his progeny went off and fornicated with animals, producing black people, who are subhuman beings.

The racist focus of the Christian Identity Movement would certainly disqualify its claim to being Christian. In a similarly ironic fashion, its appropriation of 'Israel' is laden with anti-Semitism, such that Jews are not only seen as the living embodiment of the Devil but, by falsely claiming to be Israelites, they are guilty of deception. Not surprisingly, one finds Christian Identity groups accusing Jews of every kind of conspiracy, including that of the Holocaust, which

---

13. Joseph Fielding Smith, *Doctrines of Salvation,* Volume 1, 66–67.

it claims did not happen. 'Conspiracy' is the sort of view which is like flypaper—just about anything in the air will stick to it. The issue of gun control is a good example. Many of these Identity groups are militant 'survivalists', also known as 'patriots'; that is, they store arms and maintain militia groups and training camps, they don't pay taxes, and they also refuse to use driving licences or social security numbers (in America and Canada) because they do not trust their governments, which they believe have been co- opted by the Devil and his minions (i.e., Jews, Asians, blacks, communists, etc . . .).[14]

Gun-control measures, like those imposed by the federal Government of Australia after the Port Arthur Massacre in 1996, are believed to be the result of a conspiracy between 'Howard and his Jews', who planned the massacre, convicted an innocent white boy of the crime, and then used it to justify the disarming of the populace.[15] In America, the infamous 'Oklahoma Bomber', Timothy McVeigh, made a bomb weighing nearly 5000 pounds (2270 kilograms), detonated it, and killed 168 people and injured six in the Alfred P. Murrah Federal Building in 1995. A year later, Eric Rudolph, known as the Atlanta Bomber, planted a dynamite-and-nail bomb at the Centennial Olympic Park in Atlanta, killing a woman spectator and injuring 111 others. On the run for five years, in 2005 he confessed to three other bombings, which targeted homosexuals, abortionists and socialists. His stated reason for the Olympic Park bombing—which is reproduced on the internet—attracted some extreme right-wing Christian supporters, who remain among his staunchest advocates. Indeed, one of them called for an Eric Rudolph Day in preference to Martin Luther King Day. This is how Rudolph justified his actions in Atlanta:

> Even though the conception and purpose of the so-called Olympic movement is to promote the values of global socialism, as perfectly expressed in the song 'Imagine' by John Lennon, which was the theme of the 1996 Games even though the purpose of the Olympics is to promote these despicable

---

14. Richard Abanes, *End Time Visions, the Road to Armageddon?* Four Walls (New York: Eight Windows, 1998), 113–14.

15. This conspiracy theory has a considerable presence on the internet, and anyone putting Port Arthur Massacre in their search engine will come up with a florid website with music: http://judicial- inc.biz/port_arthur_massacre.htm which suggests that Jews are not only behind the anti-gun lobby, but also carried out the massacre.

ideals, the purpose of the attack on July 27 was to confound, anger and embarrass the Washington government in the eyes of the world for its abominable sanctioning of abortion on demand.[16]

Both Eric Rudolph and Timothy McVeigh had an association with Christian Identity groups, although in Rudolph's case it was more an exposure to their ideas rather than formal membership. As in other instances of terrorism, his statement is a lesson in the power of extremist propaganda and arcane beliefs to generate a pathological sense of superiority. One need not be a card-carrying member of any such organisation to be profoundly influenced by its worldview, its prejudices, and its agenda.

The migration of the ten lost tribes has been translated into a very long and convoluted mythology by a variety of groups which have appropriated the original slight account from 2 Kings in the Bible and applied it to themselves with fanciful embellishments. At roughly the same time that Joseph Smith was promoting his version of American religious history, another parallel movement called British Israelism— also known as Anglo Israelism—propounded the view that the lost tribes went to England first and became the Celts. One of the leading figures in this school of thought was John Wilson, whose *Lectures on our Israelitish Origin*, published in 1840, proclaimed that the British were actually God's chosen people and the true descendants of the ten lost tribes. A variety of churches and organisations are still dedicated to upholding this view, including the Orange Street Congregational Church, just off Charing Cross Road in the heart of London. Its official beliefs demonstrate how elastic the concept of the lost tribes of Israel can be and how it cavalierly dismisses even the most rudimentary knowledge of European history:

> The Lost Tribes of Israel: History tells us their migrations [were] from Assyria via Southern Russia through Europe under a variety of names such as Cimmerians, Goths, Angles, Jutes, Saxons, Danes, Vikings and Normans. We believe the true descendants of Israel are in the Celto- Anglo-Saxon, Scandinavian, Germanic and Dutch/Holland peoples in

---

16. Statement of Eric Rudolph, 13 April 2005: http://en.wikisource.org/wiki/Statement_of_Eric_Rudolph

Australia, Canada, South Africa, New Zealand and America (A Great Nation) and the British Isles (A Great Nation and Company of Nations). We believe the Royal Family is directly descended from the Line of David.[17]

The royal family might be delighted to claim such a messianic pedigree, but like all these reconstructions and retellings of the 'true fate' of the ten lost tribes of Israel, it is based on the assumption that they were actually lost, having migrated far from their original Middle Eastern environs. But such is not necessarily the case, as evidenced by the records of Jews who accidentally discovered or went in search of their ancestors. This was a considerably rarer phenomenon than the Christian preoccupation with the exiled Israelites, which inspired overseas exploration and dreams of untold wealth. The Solomon Islands, for example, were named after the great Jewish king of the Bible because, in the sixteenth century, the Spanish explorer Alvaro de Mandana believed that there was a large island in the Pacific which was rich in gold and inhabited by Jews, being the place of King Solomon's famed goldmine, Ophir, which no one has yet located. The tantalizing Biblical passages are found in 1 Kings 9:28: 'They sailed to Ophir and brought back 420 talents of gold, which they delivered to King Solomon' and 1 Kings 10:22: 'The king had a fleet of trading ships at Tarshish with the fleet of Hiram. Once every three years it returned carrying gold, silver and ivory, apes and peacocks'. The Dutch East India Company sent Abel Tasman (1603–59) to investigate the rumours, but he discovered Tasmania instead!

Jews, on the other hand, did not engage in such costly endeavours to discover the remnants of their past, but they did at times send out emissaries to inquire about rumoured communities of their kinsmen. Thus have individual travelers and traders recorded some intriguing accounts of their encounters with isolated Jewish communities, which have been handed on through the centuries and in the modern period have found their way into various collections, such as Elkan Nathan Adler's *Jewish Travellers in the Middle Ages*, originally published in 1930 and based on J.D. Einstein's *Treasure of Travel* published in 1926. Not all include references to lost tribes, but those that do usually begin with the writings of Eldad the Danite.

---

17. Orange Street Congregational Church official website: http://www.orange-street-church.org/text/beliefs.htm

Eldad, who lived in the ninth century, was of the northern tribe of Dan, one of those which are said to be among the lost ten tribes (called Dan, Asher, Naphtali, Ephraim, Manasseh, Gad, Simeon, Reuben, Issichar and Zevulun). Eldad led a very adventurous life, travelling across the Indian Ocean, around the Persian Gulf to Baghdad, then north to Africa and even to Spain. He was shipwrecked, fell into the hands of cannibals, and was sold as a slave to a Jew, who then freed him. Eldad learnt about the Jews in Spain who were thriving under the relatively benign Islamic rulers of the Ummayad Caliphate (756–1031). He wrote the Jews of Spain a letter in which he detailed the history of his tribe and the location and character of the other Israelite tribes in the region.

In short, according to the letter, the tribe of Dan had emigrated to Ethiopia before the Assyrians had had a chance to exile them, as they were unwilling to be marshalled into the foreigners' army to fight against their own brethren to the south. The Danites, who were reputedly fearsome warriors descended from Samson, picked up their belongings and crossed the southern Nile to the land of Ethiopia, where they fought the inhabitants and settled down to an agrarian existence. Eldad recounted the captivity of the Reubenites and the Gadites and half the tribe of Manasseh, then of Asher and Naphtali, but added that the tribes of Gad, Asher and Naphtali also migrated to Ethiopia, where they had large encampments, herds of sheep, farms, and precious metals and jewels. Together the four tribes, in succession and all year around, went to war to protect their territory, while a separate tribe 'of Moses our teacher' (the Levites) lived in finery and upheld the law. Eldad was keen to impress upon his readers in Spain that the Danites were observant Jews: 'And they can speak only the Holy tongue and they all take ritual baths and never swear. They cry out against him that takes the name of God in vain, and say that by the sin of cursing, your sons would die young.'[18]

Not all Jews would be convinced by the authenticity or value of the account of Eldad (who, incidentally, signed his full name, which in the Biblical style included his ancestral line going back thirty-seven generations). The Jews of Kairouan (modern-day Tunisia), for example, were not persuaded of Eldad's credentials, and consulted their religious leader, the Gaon Rabbi Zemach. He assessed the

---

18. Elkan Nathan Adler, *Jewish Travellers in the Middle Ages*, (New York: Cover Publications Inc, 1930, reprinted 1987), 13.

letter and declared it on all accounts a trustworthy account of the tribes mentioned, but also noted the differences between his own community and the tribes of Israel. He explained that, among other things, these differences stemmed from the fact that the tribes of Judah and Benjamin (in the southern kingdom of Judah) 'kept the law more strictly than any' while the tribes of Israel (the northern kingdom) were inclined to forget the law and also maintained the fighting ways of their ancestors.[19]

A generation or two later, Eldad's letter was known to Hasdai ibn Shaprut (915–70), the Jewish leader, physician and diplomat in Spain under the Ummayads, who was both learned and well travelled. Perhaps it stimulated his interest to learn of communities of Jews who had achieved some kind of sovereignty or independence, which they lost after the destruction of the second temple in Jerusalem in 70 AD. In any case, Hasdai had heard from traders and travelers to the Orient that there was a community of Jews with its own king in the land of Khozar. He wrote a letter to Joseph, the king of the Khozars (Khazars), asking all the questions imaginable about how they came to be there, what they believed and practised, and from what tribe they derived. King Joseph replied in full, giving his location beyond the Black Sea, which would be today's Crimea. His kingdom was something of a confederation of tribes where, according to his own description, 'Israelites, Mohammedans, Christians and other peoples of various tongues dwell therein'.[20] Although his tribe was not readily identifiable with the Israelite tribes, he explained, 'we are descended from Japhet, through his son Togarma, [who had] ten sons [one of which is] Cusar. We are of Cusar . . .' About a fifth of the letter is devoted to a prayer to God and the traditional Jewish hope that He will return them to Zion. Quoting the Biblical prophet Malachai and the prophecy of Daniel, the King wrote: 'May God hasten the redemption of Israel, gather together the captives and the dispersed, you and I and all Israel that love His name, in the lifetime of us all.'[21]

The Kingdom of the Khazars would inspire much hope, and was immortalised in a much-loved poem 'Kuzari', by the Spanish Jewish poet and philosopher Jehuda Halevi (1080–1141), who was himself a traveler and a fervent believer in the ingathering of the exiles to

---

19. Adler, *Jewish Travellers in the Middle Ages*, 20.
20. Adler, *Jewish Travellers in the Middle Ages*, 35.
21. Adler, *Jewish Travellers in the Middle Ages*, 36.

Zion. The most famous of travelers, however, was Rabbi Benjamin of Tudela (1165-73) of Navarre in the north of Spain. From his home he travelled to Rome, to Greece, Antioch, Palestine, north to Damascus, Baghdad and Persia, then across to India and Ceylon and even China. His return journey took in Egypt and the Persian Gulf. Throughout his travels he noted the customs of the people, from Greeks to Mohammadans, people of unknown religion and, of course, his brethren the Jews. He noted the size of their communities, their synagogues, burial places and customs, as well as the kindnesses shown and the restrictions imposed on them by their various rulers. Indeed, almost everywhere Benjamin travelled he found Jews, even providing statistics of communities which often numbered in the thousands. Given their involvement in trade on the Spice Route to India and the Silk Road to China, their widespread dispersal— starting from major concentrations in the Middle East and radiating outward—is not surprising.[22]

After Benjamin of Tudela, a long line of Jewish travelers added to the growing evidence that the ten tribes of Israel had not necessarily vanished. Even if they had lost their original territories, they had at least settled in the region of the Middle East, the vast majority of them living under Muslim rule. With the establishment of the state of Israel in 1948, however, and the Suez Canal Crisis in 1956, most Jews in Muslim countries were disenfranchised, expropriated and expelled, after which they made their way to Israel under the Law of Return.[23]

One of the most amazing stories of a community 'returning home' in recent years is one which Eldad the Danite might have recognised as his own. The Jews of Ethiopia, whom Muslims and Christians called 'Falashas', meaning aliens, lived in thatched huts, ruled the mountain highlands around Lake Tana, studied the Torah (the Hebrew Bible) and were observant in the traditions of the Jews prior to the rabbis' introduction of the Talmud. They remembered a time when Jerusalem was mighty, and they prayed to return to Zion. In three separate operations, dubbed Moses, Joshua and Solomon, from 1984 to 1991 over 20,000 Ethiopian Jews were brought to Israel in what has been nothing less than a modern Exodus. Their African

---

22. Adler, *Jewish Travellers in the Middle Ages*, 36.
23. The Law of Return was enacted by the Knesset, Israel's Parliament, on 5 July 1950. It declared that the state of Israel welcomes Jews to return to their ancient homeland as citizens. It was enshrined in the Law of Citizenship in 1952.

appearance is not an issue in Israel, where Jews of all colours and cultures have regathered in the modern period.

Many other communities around the world have agitated to come to Israel, claiming that they are Jews from antiquity or, at any rate, Jews who have maintained their identity and observance of the Torah. In November 1999, a group of eighty to one hundred Maoris who consider themselves descendants of the lost tribe of Ephraim, one of the ten tribes, which they believe made its way to the South Pacific 3000 years ago, planned a visit to Israel in the hope that they would be accepted as citizens under Israel's Law of Return. Since the 1980s, Amishav, an organisation founded by Rabbi Eliahu Avihail (1932–2015), has been dedicated to investigating and verifying such claims, and, if genuine, facilitating their emigration. In 1994, Amishav brought fifty-seven members of the B'nei Israel community of Manipur, on the Burma–India border, to Israel. In October 2006, a group of 218 people belonging to the Indian B'nei Menashe tribe in the states of Manipur and Mizoram were recognised as descendants of a lost tribe of Israel, and by 2015 over 3000 Bnei Menashe arrived in Israel after an absence of thirty-seven centuries.

Were they the remains of a lost tribe that had crossed China on the Silk Road and ended up in north-east India or were they the result of later contact with Christian missionaries or Jewish travelers?[24] It is not always possible to be sure, given the dependence on oral transmission of tradition in many cases. But their dedication is really all that matters to Amishav, the organisation which has also collaborated in the conversion of 'native American Jews' in the Mexican town of Venta Prieta and in the Peruvian city of Cajamarca, both communities which had been Christian but who wished to convert to Judaism. So, in an ironic turn of events, the native Americans of the New World, who were once erroneously thought to be the lost tribes of Israel, have elected to become Jews through a formal conversion and have indeed come home to Zion.

---

24. A living community of Chinese Jews from Kaifeng, dating from the 11th Century, was discovered by the Canadian Anglican Bishop William Charles White, who wrote the definitive work on it including extensive photographs of the completely intact synagogue, *Chinese Jews: A Compilation of Matters Relating to the Jews of Kaifeng* (Toronto: University of Toronto Press, 1942). Its artefacts are held in the Royal Ontario Museum in Toronto.

Chapter 5
The Spiritual Art of Medicine

> Ceremonies are codes. They are an alchemical patterning
> of holistic intelligence that invoke through resonance, new
> patterns of response within the cerebral cortex, creating
> powerful brain chemistry that unfetters the intellect from
> its dogmatic space- time parameters, opening a gateway into
> a limitless continuum. Ceremonies create a brain activity
> that overrides our negative neuro-linguistic conditioning by
> activating the iconic imagery of our perfection that exists
> within the dormant data matrixes of the brain labyrinth.[1]

This breathless description by Juliet and Jiva Carter appears in a
pamphlet advertising 'The Reconnection of the Human Bio-circuitry
Through Coded Ceremony'. Despite the scientific- sounding language,
it would make no sense to a medical scientist, but it would definitely
ring bells with anyone who believes in the power of the mind to
change the body. Promises of 'limitless continuum' and 'perfection'
might also appeal to the yearning for the key to immortality, while
claims to 'reconnect electromagnetically to the SOURCE' make
human regeneration seem as easy as plugging a device into the mains.
In fact, the Carters demonstrate the way in which the latest scientific
breakthroughs, such as the mapping of the human genome, can be
absorbed into spiritual practices that claim to halt the course of aging
and death.

---

1. Brochure by Juliet and Jiva Carter, 'The Covenant', an excerpt from 'The
Ceremony of Original Innocence, The Reconnection of the Human Biocircuitry
Through Coded Ceremony', Center for Harmony (Houston, Texas: Sowelu
Publishing, © 1996–99, 2000–2003).

Making the connection between the circuitry, the endocrine system, DNA activation and transcendence, we will show you that duality, disease and death are transmutable through interaction with the 'higher physics of creation' as the blueprint for a new model of existence that is indelibly written in your genetic code prior to the modification of your DNA, is resurrected THROUGH THE RECONNECTION OF CIRCUITRY.[2] This exaggerated claim is on the high end of the incredible, but the connection between the healing arts and spiritual practice has both ancient and modern credentials.

In an era dominated by science and technology an unexpected development in the medical profession is gaining ground: an interest in the healing potential of spirituality. Countless studies have been conducted since the 1970s (with the advent of Transcendental Meditation founded by Maharishi Mahesh Yogi) which have shown that the practice of meditation is an effective tool for pain management. Meditation is now among the prescribed methods for the treatment of cancer patients, and in some cases it has been shown to slow down the rate at which cancer develops. As Dr Craig Hassed, a leading proponent in Australia, put it, focusing the mind in restful meditation can put the brakes on anxiety and distress, two emotions which can cause an amplification of the symptoms. It can also have specific salutary effects on a range of other conditions, such as reducing an elevated heart rate and the frequency of epileptic seizures.[3] A senior lecturer in Monash University's Department of Community Medicine and Practice who has also taught at Harvard Medical School, Dr Hassed is keen to broaden the medical profession's awareness of the spiritual component of mental and physical health, which he believes are profoundly related. He also cites the now large body of studies which have shown that active religious faith correlates to a lower-than-average incidence of disease, substance abuse and depression. Hassed is one voice in a small but confident chorus of medical researchers who are prepared to make the remarkable claim that, in a sense, 'the body is the shadow of the soul'. This 'holistic' view of health questions the conviction, still resolutely held by the medical

---

2. Brochure by Juliet and Jiva Carter, 'The Covenant'.
3. Craig Hassed, *New Frontiers of Medicine: The Body as the Shadow of the Soul*, (Melbourne: Hill of Content, Melbourne, 2000), 42–43.

profession, that modern science alone will be able to solve the health problems that humanity faces, including mortality itself.

The leading representative of the exclusively scientific view is Richard Dawkins, British author of *The Selfish Gene* and unrepentant opponent of all things religious.[4] His 2006 two-part BBC television program on the catastrophic effects of religion on civilisation was unsubtly called *The Root of All Evil*. Dawkins presents religion as a brainless, extremist, anti-humanist and violent enterprise, which has left murder and mayhem wherever it has insinuated itself into human affairs—which is everywhere, since no human culture has done without it (except, of course, in pockets of the free world where people like him, with no taste for faith practices, can champion the cause of life without religion). There is little doubt that the prospect of Buddhist monks or Christian contemplatives invading medical settings or educational facilities would upset Dawkins. Indeed, he would see it as a throwback to the time when medicine was part scientific knowledge and part hocus-pocus.

In fact, it was only in the nineteenth century that medical science emerged from a world that was heavily reliant on harebrained theories about the origin of illness, accompanied by positively brutal practices of doctors who had not the slightest idea about hygiene nor any knowledge of the natural healing processes of the body. Physicians administering lethal ointments made from toxic metals such as mercury, or pouring boiling oil onto an open wound, would as easily prevent the gashed skin from naturally closing up as the surgeon's unwashed instruments would spread disease. In most cases it would have been wiser to stay away from doctors and steer clear of hospitals, which were little more than halfway houses to the morgue.

The influence of the mind on the body however was coming into vogue in the nineteenth century, when temporary relief came by way of the truly impressive art of hypnosis, where the hysterical underpinning of some symptoms was demonstrated by the French Professor of Medicine at the University of Paris, Jean-Martin Charcot (1825–93). Sigmund Freud's early exposure to him in 1885 set the Viennese doctor on the road to psychoanalysis, although he differed from Charcot, who believed that his patients were suffering from

---

4. Richard Dawkins, *The Selfish Gene*, 2nd edition (Oxford: Oxford University Press, 1989).

neurological disorders. In any case, the real cure for many serious illnesses would not be made by men swinging a pocket watch but by medical researchers who realised that the living organisms revealed under the microscope were potentially deadly. (Bacteria were first described in a bemused fashion in 1683 when the Dutch scientist Antony van Leeuwenhoek discovered these 'animalacules' in his tooth tartar and saliva.) When penicillin was discovered by Alexander Fleming in 1928, and used successfully in 1942 by Bumstead and Hess as a powerful antibiotic to treat humans, it looked as if medicine would never look back to the past, when it was the preserve of priests, magi and witchdoctors.

Medicine is an ancient art, and its relationship to religion was for the most part a natural consequence of the belief that life was sacred; indeed, a gift of divine consequence. It would stand to reason, therefore, that when the body's workings went woefully awry, the reason might have to do with a lack of faith or the intrusions of the Devil. This meant that healing was significant in two ways: first, the person who had the power to heal must have been invested with an extraordinary measure of divinity; and second, for those who suffered from an ailment, a healing performed on them not only had the power to heal their body, but also to restore their faith. One need only be reminded that the most important signs that Jesus gave to his followers to convince them that he was the divinely anointed one, the messiah foretold in the Scriptures, was to heal the sick and bring the dead back to life. Whether he was curing leprosy, congenital blindness, deafness, paralysis, fever, shriveled limbs, muteness, a haemorrhage of blood, or madness, not to mention raising the dead, Jesus' power to heal was legendary—although, it must be said, it was not in itself unique.

There were many healers in antiquity, as there are today, and a younger contemporary of Jesus from Galilee is remembered in the ancient Jewish book of rabbinic law, the Talmud. Rabbi Hanina ben Dosa was revered for healing by prayer alone, and at a distance, the mortally ill son of his teacher, Rabbi Gamaliel the Elder. He repeated the miracle of healing on the son of his other master, Yohanan ben Zakkai. In yet another case, Hanina, whose concentration during prayer was phenomenal, was not even disturbed when a snake bit him. On the contrary, on that occasion the snake, which had been injuring others, died after it bit the rabbi. This man of such holiness, with talents which included controlling evil spirits and bringing forth rain in a drought, nonetheless lived in the utmost poverty. Why? Because his piety and humble circumstances were the very qualities which earned him the God-given power over life and death.[5]

The connection between devout faith and wellbeing had been spelt out in the Bible as indeed its opposite was also made clear in the opening chapters of the Book of Genesis, where mortality itself was introduced into the Garden of Eden as a result of Adam and Eve's disobedience. Having lost their chance at immortality, humans of Aegae still had the possibility of long life. Hence the ancient Israelites were told in Exodus 15:26 that disease would be visited on the faithless, whereas obedience to God's laws would mean escape from pestilence and illness: 'If you listen carefully to the voice of the Lord your God and do what is right in his eyes, if you pay attention to his commands and keep all his decrees, I will not bring on you any of the diseases I brought on the Egyptians, for I am the Lord, who heals you.' The principle behind the holy man who heals and the righteous rabbi who restores health may be that they simply imitate God, who has the power to both smite and to save his people, as in Deuteronomy 32:39: 'See now that I myself am He! There is no god besides me. *I put to death and I bring to life, I have wounded and I will heal* [my emphasis], and no one can deliver out of my hand.' The belief that God healed the sick was well established in the liturgy of the Israelites, who gathered to sing praises to the Lord, and recounted his powers, as in Psalm 103:1–5:

> Praise the Lord, O my soul; all my inmost being, praise his holy name. Praise the Lord, O my soul, and forget not all his benefits—who forgives all your sins and heals all your diseases, who redeems your life from the pit and crowns you with love and compassion, who satisfies your desires with good things so that your youth is renewed like the eagle's [my emphasis].

Perhaps the most compelling healer around the time of Jesus was the pagan Apollonius of Tyana, born around 4 BC at Tyana in Cappadocia, about 100 kilometres north of Tarsus, the birthplace of the apostle Paul. At the age of sixteen Apollonius, who was from a

---

5. Mishnah Berakot 5:5 (compare with Matthew 8: 5–13). Babylonian Talmud, Berakot 33a.

wealthy family, started his training in philosophy. He was a devotee of a sect which preserved the teachings of the sixth-century BC Greek mathematician and mystic Pythagoras, whose priests were adept at magic, the laying on of hands and the interpretation of dreams. Apollonius also studied with the Epicurean philosopher Euxenes of Aegae Flavius for sixteen years, but rebelled against his indulgent ethic and adopted a life of austerity, vegetarianism and travel in search of knowledge, all characteristics of Pythagoras. Apollonius' demonstrations of healing and clairvoyance impressed his teacher, and were celebrated wherever he went. But it was his meeting of the Buddhist monk Zarmaros of Bargosa which would lead him to India in search of greater knowledge.

Zarmaros (who is remembered by several ancient writers, including the first-century Strabo[6] and the third-century Dio Cassius[7]) had visited the West in 20 BC but was reputedly so disappointed with the Greek secret society of Eleusis and its Mysteries (the initiation ceremonies associated with the worship of Demeter, the goddess of life, agriculture and fertility, and her daughter Persephone, a symbol of regeneration) that he demonstrated to its adherents the direct path to God—self-sacrifice, also known as letting go of the body. Zarmaros set himself alight by oiling his body and leaping onto a pyre on the altar of Demeter. His act of self- immolation ignited a fascination for Buddhism in Ancient Greece, and in particular its denial of death, which Apollonius incorporated into his teaching, as is evident in a story about his reputed ability to raise the dead: during a funeral procession for the young daughter of a Roman family, Apollonius is said to have cried out, 'Set down the bier and I will dry the tears being shed for this maid', and the young girl was immediately revived. He explained his miracle thus:

> There is no death of anything save in appearance. That which passes over from essence to nature seems to be birth, and what passes over from nature to essence seems to be death. Nothing really is originated, and nothing ever perishes, but only now comes into sight and now vanishes. It appears by reason of the density of matter, and disappears by reason of

6. Strabo, *Geography*, Book XV: On India: 4, 73 (New York: Loeb Classical Library, 1930).
7. Dio Cassius, *Roman History* (New York: Loeb Classical Library, 1924), 54.9.10.

the tenuity of essence. But it is always, the same, differing only in motion and condition.[8]

The concepts of no origination and no perishing (no birth and no death) are central and distinctive to Buddhism, and although it is difficult to see how it could account for Apollonius' ability to revive a dead person, it is nonetheless a sign of his willingness to adopt novel ideas into his philosophy.

It is probable that the full account of the travels of Apollonius and his disciple Damis, who recorded their encounters with sages, Brahmans, kings, monks and even an elephant rider, or *mahout*, was highly romanticised. In any case, Damis' diary was a key source of information for Flavius Philostratus, who published his life of the magus in 217 AD, and was keen to defend him from accusations of malevolence and sorcery levelled by rival philosophers and the early Church fathers, such as the Christian theologian Origen of Alexandria. It may be that the fabled healing powers of Apollonius bore a certain resemblance to those of his contemporary Jesus, and may explain why Christian historians failed to mention him. This was suggested by James Loeb in the preface of the Loeb Classical Library edition of Apollonius' biography published in 1912 and was echoed by the translator F.C. Conybeare in his introduction.[9]

Certainly, Apollonius was a striking figure. He wore only linen and nothing made of animal hide, did not cut his hair, walked barefoot, and refused to eat meat sacrificed to idols. He even took a vow of silence for five years, which might have been in imitation of the *sadhus* (ascetic renunciates or holy men) whom he saw in India. Apollonius' admirers and followers would have had no trouble arguing which holy man—Apollonius or Jesus—was the more triumphant over death. They would have known that *their* long-haired holy man, who did not let a razor touch his face, was, like Jesus, persecuted by the Roman Emperor Nero, and later even imprisoned by the Emperor Domitian. However, he did not die in ignominy on a cross at the age of thirty-three, but lived to the noble age of one hundred. For a society that believed in the corporeal proofs of pure faith, it was obvious that Apollonius knew the true path to immortality—or ripe old age, at least.

---

8. Quoted in Alan Baker, *The Wizard: A Secret History* (London: Ebury Press, 2003), 41.
9. F C Conybeare's was only the second translation of Philostratus' work to appear in English; the first was by an Irishman, the Reverend E Berwick, in 1811.

Had Apollonius travelled as far as China, he would have encountered Daoism, or 'the way of virtue', as it is described in the philosophy's central text, the Daodejing, and its sages would have introduced him to the secrets of immortality. Indeed, the Daoist way of life, which legend states was expounded by Lao Tzu around the third or fourth century BC, assumes the existence of a natural force or power (the *Dao* also spelled *Tao*) which pervades the cosmos and embraces all life.[10]

To realise 'the way' is to focus on what is not apparent, what is absent, and what is the opposite of the expected norm.

> Look for it, and it cannot be seen; Listen for it and it cannot be heard; But use it and it will never run dry![11]

Daoism offers an experience of harmony with all of creation by being in tune with the life force, the *Dao,* that never rests as it pulses through the cosmos, and is apparent in the five elements, earth, air, water, fire and metal. The ideal outlook of a Daoist was a nimble awareness of the constant flow of change, and a response to it that did not create misfortune. This harmonious relationship to the universe was produced by an attitude of mind, which the Daodejing aimed to cultivate in its readers. Water was the symbol of the sage: it filled the depressions before moving onward, and flowed around obstacles rather than through them. Not surprisingly, Daoists believed that such an attitude would bring long life and even, in some cases, immortality. Indeed, among the Chinese, the Daoists were the foremost promoters of the occult notion of becoming immortal and developed a vast number of treatises and manuals on alchemy, detailing recipes of metallic, chemical and herbal substances that would achieve it. The aim was to join the mythical immortals, one of whom was Lao Tzu himself, who were said to dwell beyond the mountains. A prolific tradition of Daoist paintings depicted these immortals in their mountain paradise, in valleys and caves, just beyond the peach trees. Although it is easier to believe in the ethereal promise of the enchanting paintings than in the turgid alchemical treatises, in practical terms Daoism's theories of body chemistry generated a system of Chinese medicine which is still used today.

---

10. Philip Ivanhoe, *Daodejing of Laotzi* (Indianopolis: Hackett Publishing Co, 2003), chapter 35.
11. Ivanhoe, *Daodejing of Laotzi,* chapter 35.

The rustic appearance of Lao Tzu, the dishevelled look of Apollonius, the extreme poverty of Hanina ben Dosa, and the poor and the outcast friends of Jesus outwardly link these spiritual wonder-workers. It was little different for the bedraggled and unkempt Paracelsus, the man who, like the Daoist, was greatly interested in alchemy as the key to long life. A physician who broke with tradition and a wandering mystic, Paracelsus was considered the reformer of medicine on a par with his contemporary, the reformer of the Church, Martin Luther. Born in 1493 in the Swiss monastery town of Einsieldeln, Theophrastus Philippus Aureolus Bombastus von Hohenheim was the son of a doctor and destined to follow his father's profession—though in a most unorthodox fashion. He was schooled in Italy and took on the sobriquet Paracelsus, probably in homage to the first-century Roman healer Aurelius Cornelius Celsus. His mother died when he was young, and the absence of maternal love

FAMOSO DOCTOR PARESELSV

Unorthodox reformer of medicine, Paracelsus, was an itinerant doctor who gave up his profession in 1529 to become a wandering lay preacher for a number of years.

led the psychologist Carl Jung to speculate in an essay on Paracelsus, that the young man would find a surrogate in Mother Church and Mother Nature. Certainly, like Giordano Bruno after him, Paracelsus found no contradiction between the growing knowledge of the natural world and God's purpose. But unlike Bruno, his medical preoccupations were of more practical use to the public than Bruno's theories of the universe or his techniques of memory enhancement.

Because Paracelsus treated large numbers of people, some of them successfully, he would always have his protectors, no matter how badly he behaved in local inns or how little care he devoted to his personal appearance. On the other hand, his writings were often such a mixture of great insight and thoroughly opaque eccentricities and ravings that he was probably beyond the understanding of many. 'Strange, new, amazing, unheard of, they say are my new physics, my meteorics, my theory, my practice,' says Paracelsus of his critical reception.[12] Paracelsus was an original thinker, a prolific writer and an itinerant doctor who travelled far and wide in search of knowledge. He was also a man whose profound belief in God compelled him to interrupt his full-time medical occupation between 1529–35 and wander Europe as a lay preacher.[13] Not that his faith was ever far from his healing art, since it provided the foundation from which he set out to know the world that the Great Designer had conceived and to learn to heal the human body that the Creator had made in His own image. By his own account, he travelled all around Europe, and as far as Cairo and Jerusalem.

> The arts are not all confined within one's fatherland, but they are distributed over the whole world ... If we would go to God, we must go to Him, for He says: Come to me. Now since this is so, we must go after what we want. Thus it follows: if a man desires to see a person, to see a country, to see a city, to know these same places and customs, the nature of heaven and the elements, he must go after them ... and competently enquire; and when things go best, move on to further experiences.[14]

---

12. Philip Ball, *The Devil's Doctor: Paracelsus and the World of Renaissance Magic and Science* (London: William Heinemann, 2006), 285.
13. Ball, *The Devil's Doctor: Paracelsus and the World of Renaissance Magic and Science,* 109.
14. J Jacobi, *Paracelsus: Selected Writings* (Princeton: Princeton University Press, 1979), 143.

Paracelsus thought there was nothing in the world that did not bear examination because medical cures were different in various locales due to their unique astral influences, climate and geography. Hence, 'The physician should take this into account and know it, and accordingly he should also be cosmographer and geographer, well versed in these disciplines'.[15] If this sounds like a fair description of a Renaissance man, it is also close to the operating assumption of the physician-as-natural scientist, or medical anthropologist, for whom gathering empirical evidence knows no limits. In a similar vein, he was convinced that there was no part of the human body that was out of bounds for the physician. Thus, Paracelsus could appear outrageous to his Renaissance colleagues, who were more like academics, studying books and theories but not sullying themselves by actually handling the patient—that was left to the barber or the butcher, who usually doubled as a surgeon. When Paracelsus lectured physicians in Basel, Switzerland, for instance, and presented as his main exhibit a plate of human excrement, his audience took offence and fled. But Paracelsus was only being consistent with his belief that the body hid in its natural processes the divine essence of life, a power and an energy that was 'occult' in that it was hidden within the putrefied matter, not visible to the naked eye. 'Decay is the midwife of very great things', he wrote. 'It brings about the birth and rebirth of forms a thousand times improved. This is the highest mystery of God.'[16] Indeed, he hypothesised, the opposite of decay spelt sickness. Paracelsus identified many illnesses which he believed resulted from the body's inability to break down certain substances, leaving a hardened 'refuse' that he named tartar, which caused gallstones, gout and arthritis.[17]

If Paracelsus had met a Chinese doctor employing the principles of Daoism, he might have found the tenets familiar. The Swiss doctor wrote in his *Opus paramirum* that the universe was permeated by an invisible force that was responsible for the inherent productivity of the natural world. He called this force *Iliaster* (probably from the Greek *hyle*, for matter, and the Latin *astrum*, referring to the stars and destiny,

---

15. N Goodrick-Clarke, translator, *Paracelsus: Essential Readings* (Berkeley: North Atlantic Books, 1999), 74.
16. J Jacobi, *Paracelsus: Selected Writings*, 144.
17. N Goodrick-Clarke, *Paracelsus: Essential Readings*, 90.

according to biographer Philip Ball[18]), conjecturing that it conspired with the alchemical nature of matter, resulting in the myriad forms that matter always seeks to become. Compare a similar concept of the *Dao*, described in the poetic language of the *Tao Te Ching*.

> The Tao is like a well:
> used but never used up.
> It is like the eternal void:
> filled with infinite possibilities.
> It is hidden but always present.[19]

Paracelsus, by contrast, did not have a poetical cast of mind, nor was he attempting to express a mystical truth; his words are those of a primitive scientist who relied primarily on his imagination to conceive of a set of mechanical relationships that would explain the way the world worked. Invisible though these interactions were, Paracelsus hoped to make them explicit and thereby save people from indulging in fruitless superstition and useless remedies.

> You have seen how natural bodies, through their own natural forces, cause many things [deemed] miraculous among the vulgar. Many have interpreted these effects as the work of Saints; others have ascribed them to the Devil; one has called them sorcery, others witchcraft, and all have entertained superstitious beliefs and paganism. I have shown what to think of all things.[20]

Yet Paracelsus's belief that a life force infused all of nature, from the furthest stars to man himself, was also entirely supported by a profoundly religious outlook, and was expressed in language remarkably similar to that of the revered fourteenth-century German preacher and mystic Meister Eckhart, whose writings or ideas Paracelsus surely would have known. As a physician bringing remedies to the sick, Paracelsus dispelled the notion that their illness was due to sinful corruption or the work of Satan, but instead

---

18. Ball, *The Devil's Doctor*, 285.
19. Stephen Mitchell, translator, *The Tao Te Ching* (San Francisco: Harper Collins, 1988), verse 4.
20. HM Pachter, *Paracelsus: Magic into Science* (New York: Henry Schuman, 1951), 217, cited in Philip Ball, *The Devil's Doctor: Paracelsus and the World of Renaissance Magic and Science*, 284.

delivered a message of great hope with these words about the God who dwells within:

> For God, who is in heaven, is in man. Where else can heaven be, if not in man? As we need it, it must be within us. Therefore it knows our prayer even before we have uttered it, for it is closer to our hearts than to our words, God made his heaven in man great and beautiful, noble and good.[21]

No wonder the body, imbued with God, had the power to heal itself. Although Paracelsus was remarkable for breaking with traditional scholastic approaches to medicine and insisting upon empirical investigation, he was certainly not scientific in the sense in which that term is used today, where the empirical scientist's task is to disprove a theory in order to test its veracity and amend it. He reflected a more medieval approach, simply amassing 'proofs' or confirmations of his assertions (and ignoring conflicting evidence). But the attraction of his approach to healing is clear, analogous as it is to salvation itself. Just as the latter was the result of man's union with God, so the promise of a healthy body and long life was founded on communing with the life force of the universe.

Paracelsus was a significant influence on many nineteenth-century thinkers, who saw in him the perfect example of theosophy—the spiritual science that combined an independent pursuit of knowledge of the world—with the occult, alchemy and a divine force. Helena Blavatsky (1831–91), who with Henry Steel Olcott founded the modern spiritual movement Theosophy in New York in 1875 (see chapter 7), described her early education reading the books in her grandparents' library, which contained hundreds of volumes on alchemy, magic and the occult. 'I had read them with keenest interest before the age of fifteen . . . Soon, neither Paracelsus, Kunrath, nor C Agrippa would have anything to teach me.'[22] She even attributed to these thinkers the combination of the masculine and feminine in the manner of the Eastern (Daoist) principle of *yin* and *yang*. She was well aware of her indebtedness to Paracelsus (and other writers

---

21. Philip Ball, *The Devil's Doctor: Paracelsus and the World of Renaissance Magic and Science,* 285.
22. Sylvia Cranston, *HPB: The Extraordinary Life and Influence of Helena Blavatsky: Founder of the Modern Theosophical Movement* (Path Publishing House, 1993), 31.

on the occult, like Eliphas Levi), and anticipated the accusation that she had plagiarised his ideas in her tome, *The Secret Doctrine*.[23] In her defence, she admitted that the ideas were not original to her, but that she had merely brought them together: 'I have made only a nosegay of culled flowers and have brought nothing of my own but the string that ties them.'

One of these ideas was that the physical body contains within it the astral body, a storehouse of energy which is released upon death. This is an idea that she could easily have picked up from the study of Indian Tantra, which posits the physical body and the subtle body, the latter being an imagined metaphysical parallel body, which is connected to the cosmos through divine energy, or *Shakti*. During Blavatsky's time, the Bavarian-born American doctor Franz Hartmann, who was on the Board of Control of the Theosophical Society in Adyar, India, became a noted author on metaphysical subjects, especially renowned for his 1887 book on Paracelsus. When Madame Blavatsky died there was said to be a huge noise, like a wall crumbling or plates smashing (an obvious correspondence to the sound of an earthquake upon the death of Jesus, recounted in the Gospel of Matthew 27:51). Dr Hartmann explained the noise in terms of Paracelsus' idea that the astral body (which, unlike the visible body, interacts with the occult forces of nature), when released after death, can produce this noise because a huge rush of energy is unleashed.[24]

Another influential nineteenth-century idea drawing on Paracelsus' notion of the body's life force was mesmerism, named after German- born and Viennese-trained Franz Anton Mesmer (1734–1815). Mesmer thought of the life force as the natural flow of ether or magnetic fluid that surrounds and suffuses a body which, when subject to blockages, could result in a variety of ailments. He believed that the unimpeded flow of ethereal fluids could be affected by the application of magnets, which he designed in ever more elaborate constructions, including a series of magnets immersed in water that was then sprayed onto large numbers of people, usually women,

23. Helena Petrovna Blavatsky, *The Secret Doctrine*, 2 volumes, 1988 facsimile reprint, The Theosophy Company, Los Angeles, California; 1974, Pasadena, California Theosophical University Press; 1988 reprint, Boris de Zirkoff, editor (Adyar Madras, India, Theosophical Publishing House, 1978), Part 1, xlv.

24. Cranston, *HPB: The Extraordinary Life and Influence of Helena Blavatsky: Founder of the Modern Theosophical Movement*, 409.

holding hands in a circle (perhaps this was the origin of the wet T-shirt parade!). He went a step further when he became convinced that his own body contained a magnetic force, which, for instance, caused an increased blood flow in a woman when he drew near to her. Believing his touch could heal people, Mesmer began instructing his mainly female patients to wear only loose smocks while he applied his healing hands to their breasts, thighs and elsewhere. Whatever complaints his patients had prior to their healing sessions with Franz, they soon forgot them and he was proclaimed a miracle worker. Incidentally, when the newly discovered invisible force of electricity seemed to support Mesmer's ideas, it was used to provide stimulus for women suffering from hysteria. Ian Bersten, the Australian historian of coffee, entrepreneur and foremost collector of antique coffee machines and related home devices, has among his more exotic items several electric orgasm machines which, following Mesmer's discoveries, made their way into the doctor's office specifically for the treatment of women at the turn of the twentieth century.[25]

Mesmer was a seminal figure in the art of charismatic healing, also called 'animal magnetism', a term in which 'animal' was perhaps more apt than 'magnetism'. Setting the salacious aspects aside, Mesmer may have demonstrated how painful complaints and bodily distress find relief in emotional release and a feeling of euphoria, and how physical symptoms vanish when anxiety- producing conditions, such as loneliness and sexual repression, are removed. What Mesmer ascribed to the unseen forces that pervade all space was more than likely the power of the mind to affect our physical comfort, even to stimulate those euphoria-producing endorphins which produce, among other things, that floating feeling during sex. His popularity in Paris prompted him to open a number of healing centres in Europe, run by his students, which is how the craze for mesmerism spread and gained the endorsement of satisfied customers, including Charles Dickens.[26]

Of course, Mesmer remains something of a scandalous chapter in the history of medicine because among other reasons his healing art soon expanded to include prophesying the future. This aligned him

---

25. Ian Bersten, *Coffee, Sex and Health: A History of Anti-Coffee Crusaders and Sexual Hysteria* (Sydney: Helian Books, 1999).
26. Peter Washington, *Madame Blavatsky's Baboon* (New York: Schocken Books, 1993), 16.

more with spiritualists and clairvoyants than with physicians, and led to embarrassing criticisms and exposés by sceptics like the American bunkum detector, Benjamin Franklin. But Mesmer's ethereal presence in the astral sphere was further affirmed after he died, and his students claimed to be in psychic communion with him.[27] Apart from his sexual and psychic liberties, Mesmer nonetheless was in possession of a quality that is not only a powerful therapeutic tool, but is also an undeniable advantage in the spiritual life: the ability to arouse sympathetic affinities with strangers and induce in them an altered state of consciousness. Calling himself 'a Sensitive' he was capable of putting his patients in a trance, which he called a 'crisis', and in that state manipulating their behaviour through autosuggestion. It would be the genesis of what is known today as hypnotherapy, actually perfected by a student of Mesmer, the Marquis de Puysegur (1751–1825). It is said the Marquis 'stumbled' upon the 'perfect crisis', or sleep state, when a young shepherd boy from his estate carried out his commands while the Marquis rubbed his forehead, yet had no memory of his actions when he awakened. The dubious uses of hypnotherapy are undoubtedly one of the reasons for a large degree of scepticism about it, yet it remains a widely practised treatment for people who suffer from irrational phobias and obsessive compulsive disorders. Its 'spectacle' aspect, however, has never entirely left it, as demonstrations of its powers are a popular subject of daytime TV shows.

When the acclaimed human fertility scientist Lord Robert Winston discussed the impetus for writing the book *The Story of God*,[28] also a BBC television series, he stated that 'the more scientific research that is conducted, the more we recognise that there is so much we don't know and science may never explain'.[29] While that will not stop human beings from trying to explain natural phenomena in empirical terms, it suggests why even in the field of medical science there is renewed willingness to peer over the monastery wall or through the church door for some surprising answers. There are, in fact, mounting studies to show that religious practices, such as regular prayer, meditating, ritual fasting, acts of forgiveness, charitable works, and

27. Washington, *Madame Blavatsky's Baboon*, 16.
28. Robert Winston, *The Story of God* (London: Bantam Books, 2005).
29. Interview, 'The Spirit of Things', ABC Radio National, 5 June 2006.

communal activities, are connected to physical wellbeing—the Nun Study referred to in the introduction being a case in point. (Note, this is not the same as saying that religion is always good for you; under some conditions it can be the source of great anxiety.) However, as we have seen with the advent of mesmerism and hypnotism as a cure- all in the nineteenth century, it did tap into something essentially true about the mental and emotional dimension of illness. Certainly, the faith-healing sessions that occur regularly in Pentecostal churches are seen by many as having, at the very least, a placebo effect which, though temporary, genuinely relieves the healed person of pain. What is one person's Holy Spirit is another person's rush of adrenaline, flooding the brain with endorphins and giving a feeling of almost miraculous weightlessness and energy. If Paracelsus were summing up the case, he would conclude that God's wonders are truly to be found operating in the natural world, deep within human chemistry, the workings of which were known only to God and a few humble servants like himself, who had scoured creation to learn its hidden secrets.

Chapter 6
The Snake Goddess and Mrs God

'I Am Woman', the anthem to womanhood written and sung by Helen Reddy in 1972, became an instant hit, not only for her powerful delivery, but because the message promoted an image of woman that was militant and victorious. Reddy, who died on 29 September 2020 and who is the subject of a major Hollywood film, belted out the refrain 'I am strong, I am invincible, I am woman', which today may seem dated, but for women in the 1970s it was a daring challenge to a masculine, even macho, culture. Women were at the barricades, fighting for the right to be treated on a par with men in a bid to change the cultural foundation of a society that had allocated them a subservient role. Everything was up for grabs, including God, whom feminists generally regarded as the Patriarchal Enemy. He would need to be unseated.

In fact, the struggle to overthrow God-as-patriarch was not new.[1] It had started in earnest three generations before, at the end of the nineteenth century, and the language of the suffragettes had been no less strident.

Frontal assault is one way to fight a war, but employing the element of surprise is more likely to win the battle. Who would have thought that the arcane discipline of archaeology would provide a means to victory? Yet that is what happened, both at the end of the nineteenth century and in the 1970s. In a society that respects scientific knowledge, it should come as no surprise that scholars were capable of undermining ironclad convictions through astonishing discoveries of evidence that lay buried deep in the earth. The trouble

---

1. Rachael Kohn, *The New Believers: Re-imagining God* (Sydney: Harper Collins, 2003).

is that both then and now their discoveries have been hotly contested, and the result has divided people once again into those who believe and those who do not believe in the true significance of what has been revealed.

She became known as La Parisienne: an alluring beauty on a fragment of the Campstool Fresco, Knossos, Crete, c 1400 BCE, and now in the Archaeological Museum in Herakleion. The style of her hair, her eye make-up and the elaborate 'sacral knot' of fabric around her neck suggest that she had an important cultic role, which was probably related to her femininity. Nineteenth-century Paris appeared to have discovered a version of itself in early antiquity, its claim to all things beautiful and modern given an ancient pedigree and a divine imprimatur. The qualities of 'naïve archaism and spicy modernism' that these images evoked would be given a further boost, however, with the discovery of a full-bodied figurine, which became known as the Snake Goddess. She was sexy, dressed in a beautiful hip-hugging skirt with a flounce at the ankles, bare breasted and with the curves of a well–developed young maiden. Two golden snakes wound elegantly up her outstretched ivory arms. Her wavy hair fell to her shoulders in long tendrils. In some depictions a kiss-curl adorned her forehead and her lips were painted red. This was a goddess who was worthy of the label La Parisienne, and as more damsels were discovered painted on the walls of the recently excavated palace at Knossos in Crete she became a composite of their attributes. Kenneth Lapatin, art historian and assistant curator of antiquities at the J. Paul Getty Museum in California, explains that the Minoan Snake Goddess was reproduced in art magazines and history books so many times that the archetypal Cretan beauty became as real as if she were a contemporary socialite.[2] Certainly, the ladies who affected her fashion sense wanted to be like her—and the men wished they were.

When British archaeologist Arthur Evans came into possession of the Minoan Snake Goddess in 1902, it was after eight years of digging around the island of Crete and purchasing antiquities from the locals. In fact, it was not Evans himself who found it, but his team who discovered the roughly 15-centimetre high figurine. Evans was in search of evidence of the ancient Minoan language, and although

---

2. Kenneth Lapatin, *Mysteries of the Snake Goddess* (Cambridge: Mass DeCapo Press, 2002).

The Minoan Snake Goddess, thought to be over 3500 years old and discovered on Crete in 1902, brought a scandalous allure with her when she was donated to the Boston Museum of Fine Arts in 1914.

he never found anything decipherable, what he acquired would become a sensation. The figurine, which he took to be one of the many antiquities plundered from the site of the palace at Knossos, would soon be hailed as a masterpiece—and at 3500 years old, it was years older than the many Greek artefacts that were discovered by his contemporaries in the highly competitive field of archaeology at the time. More importantly for Evans, the Minoan Snake Goddess was at least as old as the Egyptian and Mesopotamian artefacts, which were then all the rage. In 1914 a wealthy Bostonian, Henrietta Fitz, bought the figurine for US$950 and donated it to the Boston Museum of Fine Arts, where it became their most celebrated antiquity.

Our Lady of Crete, as I like to think of her, inspired fashion, art, even performance, as ladies of leisure indulged their artful fantasies and re-created dramatic scenes from antiquity in their salons and gardens. She not only arrived on the scene during a wave of interest in Greek antiquities, but she also struck a chord with her bold and risqué stance, one that mirrored the attitude of modern women chafing for freedom and a place in the sun. It was as if this Minoan Snake Goddess, with her pagan charms, was poised to replace the submissive and chaste model of womanhood that Christianity had fashioned in Mary. The fact that the Snake Goddess's new home was Boston, a stronghold of Catholicism in America, certainly added to her scandalous allure. She was an affront to the draped and demure Virgin Mary, whose modesty and humility were signs of her holiness. Similar examples of the snake-handling goddess from the Knossos site were soon added to antiquities collections in Toronto, Oxford and Berlin.

The obvious sexual connotations of the Snake Goddess, seen through Christian eyes, cannot be overlooked. Snakes were invariably associated with sex, given that it was the snake in the Garden of Eden that caused Eve to eat the apple and to become aware of her sexuality. Snakes also have phallic connotations, which may well be why of all the creatures it is the snake that tempts Eve. And snakes are undeniably dangerous and perceived to be cunning creatures, slithering silently through the undergrowth and surprising their victims with a fatal bite. Snake- charmers, known throughout the ancient and modern world, were therefore considered to be more than just daring or skillful; they were thought to have a truly awesome power over creatures of evil intent. In the time of Jesus, magicians and healers were commonplace in Galilee, so it is not surprising that, after the Resurrection, among the signs and wonders Jesus promised his apostles was the ability to handle snakes: 'And these signs shall follow them who believe, in my name they shall cast out Devils, they shall speak with new tongues, they shall take up serpents' (Mk 16:17–18).

For Arthur Evans, however, the Christian view of the snake was not very helpful, and in fact was opposed to the positive role that snakes played in the cultures that fascinated him. In Ancient Crete, snakes were associated with soothsayers, while Greek myths also identified snakes with powerful figures, such as the goddess Athena, her father Zeus and the healing god of Asklepios, whose symbol of a snake wrapped around a staff, the caduceus, is still used in medical insignia today. Of the several depictions of the Snake Goddess found at Knossos, Evans chose a larger figurine—discovered in depressions that he dubbed 'Temple repositories'—to grace the front cover of his major publication, *The Palace of Minos*. This specimen, which was about twice the size of the Boston Goddess, had snakes wrapped around her hips, and buxom maternal breasts, which Evans believed distinguished her as the Goddess of Maternity or Mother Goddess.[3] Indeed, Evans was convinced that the Snake Goddess had a far wider and greater importance than a single player in a pantheon of deities or priestesses. Like others in his day, he was imbued with the late nineteenth-century mania for matriarchal cultures—perhaps a reflection of Queen Victoria's long reign—and was convinced that the ancient Minoans worshipped a Great Mother Earth Goddess, of which the Snake Goddess was a manifestation.

Of course, the shapely Snake Goddess from Crete was a far cry from the Earth Mother who became all the rage in the 1970s when Western women were in search of new icons of womanhood. As girdles and brassieres were ripped off to reveal the large frames and floppy breasts of an affluent society, another kind of goddess became the role model of choice. Not the young virgins of antiquity, but the other ancient goddess, one discovered not long after Arthur Evans found his beauty. By no stretch of the imagination could you call this goddess beautiful, yet she did have it over her Minoan counterpart in at least two ways: she was much, much older, and she had a far more extensive territorial domain. For the feminist who wanted to

---

3. *Mysteries of the Snake Goddess,* 80, 64.

throw off the sex kitten stereotype of the 1950s and 1960s and was embroiled in a fight for equality and power, it was a more primal mother goddess who became her icon, and in a curious way her squat, grotesque naked form combined raw sexuality and defiance of patriarchy in the one image.

The 'Venus' figurines of ivory, bone and stone date from 25,000 BC, and they are found from France to eastern Siberia. They are characteristic for their enlarged breasts, buttocks, hips and genitalia and their foreshortened, almost inconsequential limbs, and are believed to represent deities of fertility and abundance. Of these figures, the most renowned, although not the first to be discovered, was found in 1908 by Josef Szombathy near the town of Willendorf in Austria. As she is made from a type of limestone which is not found naturally in the region, it is believed that she was brought there from somewhere else. Be that as it may, the Willendorf Venus certainly appeared in the right place at the right time. In Europe, the ground was fertile for a Mother Goddess, and a considerable number of writers and scholars had already shown a significant interest in the notion that agriculture had been a product of matriarchal civilisation in ancient times.

One of the proponents of this idea was the Swiss anthropologist Johann Bachofen, who published *Mother Right: An Inquiry into Female Power in the Ancient World* (1861). His ideas were exploited in the early twentieth century by the charismatic personality and anarchist Otto Gross (1877–1920), who established an experimental bohemian community promoting sexual freedom and a matriarchal culture, which he believed would be the basis of an imminent revolution. Trained as a medical doctor but suffering from serious drug addiction, Gross became a cult figure who attracted intellectuals and feminists and had intimate relations with a number of his followers, including DH Lawrence's wife Frieda and her sister Else, who bore him a child.

It was with Else, too, that the renowned sociologist and comparative religionist Max Weber (1864–1920) had an affair.[4] This was no incidental liaison for the otherwise-chaste Weber. He was the German founder of 'the science of society', otherwise known as sociology, whose most important work examined the social consequences of religious beliefs, though ironically he professed none

---

4. Mervyn Bendle 'The Ark', ABC Radio National, 12 February 2006. https://en.wikipedia.org/wiki/Else_von_Richthofen.

personally. Alienated from the Church, a self- described left-wing liberal in the turbulent last years of Emperor Wilhelm II's rule, living in an unconsummated marriage to the feminist scholar Marianne Schnitger, the idealistic Weber may well have been fascinated by a 'pre-Christian' and sexually liberated cultic milieu. The personal and the political, which are never very far from scholarship, were certainly apparent in the new and burgeoning fields of sociology, anthropology and archaeology in the late nineteenth century. Then, Europeans were looking to the past and to indigenous peoples to find values and practices alternative to their own, and in similar fashion they would again a hundred years later. Just as Else, who was also a social scientist, and the others involved in the Mother Right movement found an ideological precedent for the sexual revolution of their day in the newly discovered fertility goddesses, and believed that in doing so they were re-creating an ancient, more idyllic culture, so feminist academics in the 1970s were keen to find their personal and political aspirations confirmed in them as well.

The importance of the so-called Venus of Willendorf (and many similar finds before and after her) is that her age, at 25,000 years old, was taken as proof that a matriarchal culture preceded the patriarchal values of the Bible. She was also evidence that the earlier culture was dominated by a Mother Goddess, who was mother to human beings and to animals as well. She could therefore be considered the origin of life itself. In other words, the figurines offered a glimpse into a culture that was not only a welcome alternative to the male-oriented religion that had been the foundation of Western society, but was also older and rooted in nature. The romantic view of nature as possessing a primordial order, intelligence and goodness which are more authentic than man-made civilisation has always had a powerful appeal at times of social turmoil and frustration with governments of the day. Such was the case in Germany in the period up to and during World War I, when the Mutter Recht (Mother Right) movement gained ground, and in the 1970s when the Anti-Vietnam War movement was widespread and interest in the feminist goddess arose.

A strong tendency to see war as the product of a patriarchal society which was rooted in the Bible gave proponents of a matriarchal alternative the licence to characterise it as nonviolent, nurturing and egalitarian. One of the most well-known champions of this view was comparative religionist Joseph Campbell, whose book *Hero With a Thousand Faces* (1949) became the basis of a popular television series in the 1980s, *The Power of Myth*, produced by Bill Moyers. In the foreword to Marija Gimbutas's influential book *The Language of the Goddess* (1989), Campbell wrote that goddess religion could be favourably compared to 'the manipulative systems of the patriarchal religion of the Bible', since it 'was that primordial attempt on humanity's part to live in harmony with the beauty and harmony of creation'. Gimbutas, a Lithuanian archaeologist who emigrated to America after World War II, was one of the most persuasive advocates of the view that European culture in the Neolithic Age was a harmonious matriarchal culture that knew no war and worshipped a fertility goddess whose image she found in pottery, paintings and sculptures in south-east Europe. More recently, however, serious scholars have dismissed the idealised view of goddess religion as a pure invention, relying on a highly selective reading of ancient sources that in the main do not support that view. In fact, even Hinduism, with its plethora of goddesses—such as Kali, Shakti, Lakshmi and Parvati—has not produced a society that is either peaceful or fair to women. Rather, as American scholar Wendy Doniger pointed out it had just the opposite effect.[5] Female divinities were feared because they had the deadly combination of being powerful yet seductive, with the result being that women in India were kept under lock and key, scrutinised and unfairly punished.

There is no dearth of goddesses in antiquity, and one only needs to look at the Greek and Roman pantheon, including Athena, Artemis, Hera and Demeter, to note that women could be fierce and warlike and even capable of murdering their own children. Medea is an interesting figure from Greek mythology as she has both human and divine features. The granddaughter of Helios the sun god and daughter of King Aietes, she was from Kolchis on the Black Sea. Medea is known for her love of Jason and her able attempts to help him overcome the obstacles that his uncle Pelias put in his way to prevent him from capturing the Golden Fleece. A list of Medea's exploits and powers shows that she was a formidable woman and a contender with the gods. She was a sorceress, the murderer of her half-brother Apsyrtos, and a valiant warrior who killed the enemy of the Argonauts, the bronze giant Talos. Despite all her efforts to

---

5. 'The Ark', ABC Radio National, 7 May 2006. Wendy Doniger: *Tales of Sex and Violence*, Univ of Chicago Press (1985).

prove herself worthy, her husband, Jason, fell for another woman, the Princess of Korinth. In revenge Medea caused the death of the princess and, to really add insult to injury, killed her own children, which she had borne Jason. No love lost there! Even more incredibly, Medea, who possessed the ability to resurrect dead people, did not resurrect her dead children. In other words, she was the antithesis of life-giving motherhood and family harmony. Admittedly, instances of both fratricide and infanticide are plentiful in ancient Greek and Roman culture.

If contemporary relevance could be found for Medea, she would be symbolic of the damage that spurned wives can wreak on their innocent children. But her attraction from antiquity is undoubtedly the unbridled vengeance which she wreaks with impunity on those who wrong her or stand in her way. As Helen Reddy sang triumphantly in her 1970s hit song, 'I am woman, I am invincible! . . .' But classics scholar Emma Griffiths has shown in her study of Medea[6] that she was in fact the object of considerable imaginative construction, not only by the influential ancient Greek playwright Euripides in his 431 BC play *Medea*, but also by subsequent writers who strongly suggested that it was the flaw of jealousy in her character that caused her suffering. This is precisely where feminists cry foul. They have redeemed Medea, identifying her as a victim of a patriarchal culture that cavalierly pursued women, shot the powerful arrows of Eros into their hearts, ravished them and dumped them. Medea, on the other hand, courageously fought back and emerged unscathed.

Long before the Greeks, Ancient Egypt had a plethora of goddesses. But modern women have focused their attention on the singularly powerful and most important of these figures, the Mother God known as Isis, who was worshipped all over Egypt, together with Horus, her son, and Osiris, her husband–brother. She was a sorceress and was referred to by many names, including the female version of the sun god Ra, the lady of the New Year, the maker of the sunrise and the lady of heaven. In one of the most famous Egyptian myths, Isis emerges as the symbol of undying love, for when her husband Osiris is murdered by his brother Seth, she searches everywhere for his body and, after finding it inside a pillar, sets it in a boat to bring it home. During the journey she lies on top of her dead husband and

---

6. Emma Griffiths, *Medea* (London: Routledge, 2006).

conceives a son, Horus. As Seth continues to demonstrate his hatred of Osiris, tearing his body to pieces and hiding each part in a different place, so Isis faithfully shows her love, searching for every piece of Osiris's body and building a shrine over it. There is no doubt that Isis, who is often depicted wearing horns, is the apex of goddesses, being also the lady of the solid earth, the lady of green crops, the green goddess, the lady of words, the lady of power, and also the creator of the Nile River, on which Egyptian agriculture depends. No wonder that Madame Helena Blavatsky, founder of the spiritual movement, Theosophy, and a champion of feminist religion, gave her first great tome, published in 1877, the title *Isis Unveiled*. In it she aimed to give an exposition of the new higher plane of spiritual consciousness, devoid of superstition and merged with science, which she believed was found in the great wisdom of the Egyptian goddess.

Isis was certainly popular in the late nineteenth century, thanks to the massive excavations undertaken in Egypt by British and French archaeologists. But she had a contender in the land of Sumer, a region of about 25,000 square kilometres in what is today southern Iraq. There, in around 2000 BC, thousands of clay tablets were inscribed with the songs, legends, laments and epic tales of another goddess, Inanna (known to the Semites as Ishtar). It is thought that belief in this most beloved and revered deity of the Sumerians, who was the daughter of the Moon God, Nana, goes back almost 2000 years before that to 3500 BC. Her title as Queen of Heaven and Earth, Goddess of the Morning and Evening Star, and the Goddess of Love, Grain, War, Fertility and Sexual Love make her particularly desirable as a goddess of choice. She's the closest one gets to an omnipresent goddess, which is why the stories of Inanna are many. Several of them revolve around her coming of age, gaining her sovereignty as Queen and discovering her sexuality. Others are more-worldly, to do with laws and the accoutrements of civilisation, such as garments, weapons, speech and music. She is also the focus of another spellbinding tale of sacrifice and triumph, in which she descends to the Underworld, is turned into a corpse, is revived, then re-emerges to claim back her throne as the Queen of Heaven and Earth. One Sumerian prayer proclaims:

> I say hail to the holy one who appears in the
> Heavens I say hail to the holy priestess of heaven
> I say hail to Inanna the Great Lady of Heaven

Holy torch you fill the sky with light
You brighten the day of dawn
I say hail to Inanna, Great Lady of Heaven
Awesome lady of the sky gods Crowned with great horns
You fill the earth and heavens with great light
I say hail to Inanna, the first daughter of the Moon![7]

Inanna's light did not shine forever. Like many gods, she eventually faded from view, due, in her case, to a growing patriarchy that favoured Enlil, the dominant god of the four creator gods. But Inanna is probably remembered indirectly in Iran's sumptuous celebrations of love and fertility in the New Year, No-Rooz. This spring festival goes back at least three thousand years to the Zoroastrian religion of ancient Persia, with echoes from the cult of Inanna as the goddess of love. On account of her amorous nature, many sexually explicit poems were written about her, which Diane Wolkstein has made available in a contemporary rendering that has contributed to a feminist-inspired revival of the cult of Inanna in the West.

One goddess of antiquity who was also fated to become a shadow of her former self was Asherah or 'Mrs God'. Her problem was that she was married to a man who became too powerful and revered to share the stage with his wife. Around 1250 BC, she was the wife of El (or Yahweh), the head of a pantheon of gods worshipped by the ancient Hebrews. (El is still one of the names used for God in the Hebrew Bible.) Evidence of Asherah was found at a site in Ras Shamra, Syria (once the city of Ugarit), but the earliest mention of Asherah dates back 600 years before that, to 1850 BC, when she is mentioned in texts from a site on the Upper Euphrates. In Ugarit she was the mother of the gods, referred to by Biblical archaeologist Diana Edelman as a progenitrix, or female creator.

Diana Edelman, an American archaeologist based at the University of Sheffield in England, has made a study of Asherah, and found that in the Bible she is mentioned forty-eight times, although by the time the texts were codified, after the Israelites had developed their faith into a strict monotheism, her personality had become obscured. References in the Bible nonetheless imply a divine

7. Diane Wolkstein and Samuel Noah Kramer, *Inanna: Queen of Heaven and Earth, Her Stories and Hymns for Sumer* (New York: Harper Perennial, 1983).

personage, such as when an image is made 'for the Asherah', or when there are '400 prophets of Asherah' (along with 400 prophets of the Canaanite god, Ba'al) who assemble at Jezebel's table (1 Kgs 18:19)[8]. Discoveries at Kuntillet Ajrud in the Sinai Desert confirmed the relationship between Asherah and Yahweh when Israeli archaeologist Ze'ev Meshel found inscriptions in Aramaic and Hebrew on the wall of a sanctuary situated on a trade route and dating from the ninth century BC. One inscription reads, 'May you be blessed by Yahweh of Samaria and his Asherah' and the other reads, 'May you be blessed by Yahweh of Teman and his Asherah'. This tells us that the god Yahweh was associated with shrines at specific locales. Another inscription from a burial cave refers to Yahweh of Judah and his Asherah.

Mrs God was definitely an important divinity and she gave her blessings as freely as her husband did. But their domains were different. While Yahweh's sphere of influence included things to do with territory, other tribes and sovereignty—what today we would call political issues—Asherah's domain was decidedly domestic and agricultural, having to do with fertility and the fructification of the land. The inscriptions noted above accord with an abundance of female figurines found in the area. Some of them even resemble the Venus of Willendorf, though they are not as grotesque as the Venus. The Israel Museum in Jerusalem has several of these female figurines, which some archaeologists, like Diana Edelman, readily identify as the goddess Asherah. Others, it should be noted, have referred to the little women as fertility amulets or ancient Jewish 'Barbie dolls'.[9]

But Mrs God was a casualty of political history. So long as the Jewish people of Judah, the southern kingdom, and of Israel, the northern kingdom, had their Yahweh before all other gods, they also had his Asherah, or wife. Perhaps Biblical religion would still have its divine mother and father if the Assyrians had not come 'charging down on the fold' to vanquish Israel. As discussed in chapter 4, the northern kingdom was exiled to Babylonia in 721 BC, followed by the southern kingdom about fifty years later, and it is then that the Jewish religion underwent significant changes. After the Persian conquest of Assyria,

8. 'The Ark', ABC Radio National, 2 May 2004.

9. An observation made by Lisbeth S Fried of the University of Michigan, who is more cautious about the figurines and suggests they may not have religious meaning. 'From Optimism to Doubt: Jewish Encounters with the Other', Fourteenth World Congress of Jewish Studies, Jerusalem, 31 July to 4 August 2005.

some of the Jewish exiles were allowed to return to their homeland in the sixth century BC, and they rebuilt the Temple of Jerusalem, which was completed around 515 BC. The scribes Ezra and Nehemia were keen to consolidate the community around the tradition, and they codified the Torah, or Hebrew Bible. Their work, however, reflected the development of Judaism that took place in exile, in that Yahweh became both the only God and a universal God. He was no longer the best of several gods, but absorbed them all into himself, which is signaled in the Torah by the names of several gods being attributed to the one God, Yahweh. Asherah then disappeared and Yahweh took over her role of fertility goddess. This is clear in the first books of the Bible when the matriarchs of Israel are usually depicted as barren until Yahweh opens their wombs, a role that previously would have belonged to his wife, Asherah. But by then she was reduced to a mere object, a symbol of her former self, in the Temple of Solomon and at altars, where she was carved on stone pillars or 'cult poles' of a type that were found throughout the eastern Mediterranean.

Today, there are many women who feel as if their sex has suffered the same fate as Asherah, being either reduced to dumb objects in the rituals of their religion or airbrushed out of Christian, Jewish and Muslim history. However, the disappearance of strong female divinities and the relatively recent raising of women's status has prompted a rearguard action. Women have been re- imagining and revivifying female divinities that might act as mentors and spiritual guides. They are freely and actively creating the sacred symbols, rituals and liturgical writings to celebrate these reborn goddesses.

Former Harvard academic and feminist author Carol Christ runs Goddess tours in the Greek islands, introducing participants to goddesses of the ancient Mediterranean cultures, while performing group consciousness-raising rituals to awaken the goddess within. Bearing the name Christ is a powerful thing in itself, but Carol is also revising what is understood by divine revelation. She argues that instead of a momentary event, revelation in religion has always been a continuing process, absorbing elements of previous traditions and refashioning known beliefs and practices into new acceptable forms.

Buddhism, for example, evolved from the idea of just one Buddha, who was a non-divine man, to a whole realm of quasi-divine beings called Bodhisattvas, who assist the ordinary person to gain enlightenment. One of those Bodhisattvas is Kwan Yin. She is a popular divinity to whom Buddhists pray, especially women. Carol

Christ believes that Kwan Yin can take up residence in a pantheon of similar female divinities, which includes Mary, the Mother of God, and the Shekinah, the feminine presence of God who is mentioned in the Bible and comes into her own in the vast writings of Jewish mysticism known as the Kabbalah. Carol, who developed her feminism in an ongoing conversation with Jewish women, many of them her former academic colleagues, draws on many traditions of female divinities and turns them into a living source of spirituality for contemporary women.

Carol Christ, who has lived on the Greek island of Lesbos for some years now, is nothing if not practical. Her eclectic use of divinities, including the Hindu goddess Kali, with her tongue characteristically sticking out and the sash of skulls around her middle, is aimed at empowering women. It is precisely this combination of practicality and 'will to power' that has driven the pagan goddess movement in our times, reaching its peak in contemporary witchcraft. The story of modern witchcraft, or Wicca, is well known, with the key figure in its revival being the Englishman Gerald Brosseau Gardner (1884–1964), who, after working on plantations in the Far East and later as a public servant back home, claimed to have been initiated into one of England's last surviving witches' covens. Gardner's version of his induction into an ancient religion is disputed, however, and several scholars agree that Wicca is entirely his invention, both as a modern religion and also in the sources that enable its contemporary practice.[10] Despite this now widely accepted view, Gardner's belief that witchcraft was the ancient mother of all religions, passed on in secret from pre-Christian times, remains a popular drawcard. On its revival—or advent—in the 1950s it was quickly taken up, and today the Neo-Pagan movement is comprised of a large number of groups, sometimes collectively referred to as 'earth-based spiritualities', which are more or less involved in the rituals of witchcraft. As Margot Adler summed it up in her seminal work, *Drawing Down the Moon*:

> Neo-Pagans embrace the values of spontaneity, non-authoritarianism, anarchism, pluralism, polytheism, animism, sensuality, passion, a belief in the goodness of pleasure, in religious ecstasy, and in the goodness of this world, as well as the possibility of many others. They have abandoned the

---

10. Aiden A Kelly's masterful study, *Crafting the Art of Magic: Book I, The History of Modern Magic, 1939–1964* (St Paul, Minnesota: Llewellyn Publications, 1991).

'single vision' for a view that upholds the richness of myth and symbol . . . 'Neo-Pagans,' one priestess told me, 'may differ in regard to tradition, concept of deity, and ritual forms. But all view the earth as the Great Mother who has been raped, pillaged, and plundered, who must once again be exalted and celebrated if we are to survive.'[11]

Wicca's attraction to women is, of course, the dominant role of the witch, who acts as a priestess alone or with a male priest in the rituals. There is no need for a women's ordination movement in witchcraft, which has always been a largely female domain. The witch or priestess is responsible for the practice of magic, the ceremonies, the spells and the recipes of herbal remedies, as well as teaching and initiating others into the coven. Most controversially, in some pagan groups Satan, the fallen angel who most memorably tried to corrupt the faithful Job and who reappeared as Jesus' evil adversary in the Gospels, is respected rather than reviled—perhaps in a deliberate act of defiance against Biblical religion. Satan is incorporated most often as a natural and necessary part of life rather than a menacing dark force who, like death itself, is an inevitable counterbalance to the life force. The Wiccan motto incorporates the dark side as part of the divine order: 'As above, so below.' However, its reputation for licentiousness, encapsulated in its moral motto, 'Everything is permissible so long as it does not harm anyone', has resulted in some unfortunate liaisons. In Australia, Rosaleen Norton, a witch and artist who lived in Kings Cross in Sydney in the 1950s, was the doyenne of a large circle of bohemian friends. She might never have been drawn to wider public attention had it not been for her friendship with the British-born conductor of the Sydney Symphony Symphony Orchestra, Sir Eugene Goossens.

A participant in her occult rituals, Goossens was apprehended at Sydney Airport in March 1957 when he attempted to smuggle into the country banned books, ritual masks and 1166 pornographic photographs. Fined and returned to England, his career in Australia was over, but the belief that witches engaged in dangerous sexual depravity was given a big boost. Today, Wiccans see themselves primarily as practitioners of a nature religion, with a strong pro-environment agenda. There are also many women who call themselves pagan but are not practitioners of Wicca, preferring to call their divine inspiration Gaia, the Greek name for the goddess of mother earth.

---

11. Margot Adler, *Drawing Down the Moon* (Boston: Beacon Press, 1979), 175.

Kenneth Lapatin thought he was on a roll when he finished his investigations into the Minoan Snake Goddess and discovered that all was not right with her. He learnt that the figurines which Arthur Evans and others had produced from their excavations were in fact conglomerations of fragments and re-created limbs that restorers in Greece had expertly put together. After a visit to their workshop, Evans described in detail how they worked on these figurines, a number of which were laid out in various stages of repair. But what Evans may not have realised was that these expert restorers were also capable of meeting the voracious European, English and American demand for artefacts by creating them from scratch. In fact, Lapatin was convinced by the end of his research that the Boston Museum's Snake Goddess was a fake. Surprisingly, the Boston Museum was gracious about Lapatin's announcement, perhaps because he did not single out their possession alone, but included a good many others that hold pride of place in various collections. In an unexpected twist, however, Lapatin himself started to question his own rather puritan notion of the difference between an authentic artefact and a fake one. In the end, he acknowledged that the Snake Goddess may well have had original elements, and therefore might be said to have some genuine aspects to it after all. And who is to say how much constitutes enough? Perhaps all we can expect from the urge for authenticity is to be satisfied with being a genuine fake!

Women who are appropriating ancient goddesses today are a bit like those craftsmen who reconstructed the Minoan Snake Goddess. Driven by the desires and needs of the present they are refashioning their divine guide from fragments of the ancient past. The difference is that this resuscitated goddess is a dynamic creation and not a silent object in a glass case. She is the source of meaning in women's lives and it is around her that they have created rituals and prayers, such as this one, written by Carol Christ, after the Hail Mary: 'Hail, Goddess full of grace! Blessed are you, and blessed are all the fruits of your womb. Hail, Holy Mother of All, be with us now, and in the hour of our need. Blessed be.' As of February 2020, Carol Christ announced that after 25 years on the island of Lesbos, where Sapho sang, she is moving to the island of Crete, the home of the Minoan Snake Goddess.[12]

---

12. https://feminismandreligion.com/2020/02/24/ancient-mothers-i-hear-you-calling-me-to-crete-by-carol-p-christ/

Chapter 7
The Lost Race of Giants and Aryans

The manila envelopes addressed to me in the distinctive capital letters had become familiar. I knew without opening them that they were from the person I had privately labelled 'the mad man'. Their contents still astonished me even after receiving several of them over the course of a couple of years. Carefully written in several colours of ink, the messages, all in capital letters, filled every last space of the large piece of paper, moving horizontally, vertically, around the perimeter of the page. I placed these highly unusual communiqués in the loony file and forgot about them—until to my surprise, I met their author.

A gentleman had asked to see me about an urgent matter. A tall, elegantly dressed fellow with silver hair greeted me as if he were an old friend with something on his mind. He had important information, contained in a black leather attaché case lying on the table. When he pulled out some of its contents, I recognised the distinctive writing at once. I was still trying to reconcile the eccentric contents with this well-spoken, meticulously groomed and elegantly turned-out gentleman, when I realised he was telling me about his Anglo-Asian background. Then he told me that he had communicated his very important discoveries to many world leaders, including the former American president George Bush Sr and then President Bill Clinton, former Russian President Mikhael Gorbachev and Prince Charles. He was hoping I had influence with the Australian Government, and would help him convey his findings to Prime Minister John Howard.

What was his top-secret information? He fingered the folders, showing me glimpses of satellite photos which, he explained, showed a submerged continent that revealed the true origins of the world's races. He believed their prehistory had been lost until now, and he was about to reveal the truth. There was only one sticking point; his revelation was causing great commotion among world leaders, and thus he was a marked man. Secret agents had just stolen the only other copies of his research, so he couldn't part with this folder. Tucking it back into the leather attaché case, he expressed the hope that I might be able to help him locate the other copies. In the meantime, could I help him pay for the reproduction of his precious original?

> The sons of God saw the daughters of men, that they were fair; and they took to wife such as them as they chose . . . The Nephilim were on the earth in those days; and also afterward, when the sons of God came in unto the daughters of men, and they bore children to them. These were the mighty men that were of old, the men of renown (Gen 6:2–4).

The idea of a lost race of people is at least as old as the Book of Genesis. Before the flood was sent to drown the wickedness of man, there was a mysterious race known as the Nephilim, whose stature and superhuman powers were the result of the marriage between 'the sons of God and the daughters of men'.

Many scholars today regard this story as a fragment of an earlier mythology, which produced similar tales in the ancient world, such as the Epic of Gilgamesh in pre-Israelite Sumeria around 2100 BC. The early rabbis and Greek translators of the Bible in the third century BC, who read Genesis as an account of their own prehistory, believed that 'the sons of God' were fallen angels who cohabited with humans and produced a race that was powerful but ill-fated. *Nephil* actually means 'fallen one', and is from the Hebrew root *nephal*, which means to fall. Although the rabbis understood that the extraordinary power of the Nephilim referred to their intelligence and strength, the Greek translators interpreted Nephilim as *gigentes*, giants, implying great size as well as strength. Perhaps the term was originally meant in a metaphorical sense, but by the time it was translated as 'giants' in the King James Version of the Bible, it well and truly meant huge men, like the Titans of old, of which Hercules (also known as Atlas) was one.

However this unique race was imagined, that it was the result of the union of fallen angels and humankind represented a violation in the order of creation and was thought to have generated violence and perversion among men, including fornication with the men of Sodom and Gomorrah. This angelic mischief was interpreted in both Jewish and Christian sources as an assault on the human race by the Devil, who used his minions, the fallen angels, to corrupt God's beloved creation. That is, after all, the Devil's work. But, not to be outsmarted, the Creator of the earth sent the flood to wipe all evil men, including the Nephilim, off the face of the earth.

At least, that is what was meant to happen. One thing we know about the flood, which the subsequent books of the Bible make plain, is that the evil which men did before the deluge was not wiped out after it. Clearly, the Devil and his loyal companions survived God's wrath. *So why not the Nephilim?* Given that they were half-angel and half-human, with supernatural powers and strengths, the Nephilim would have been more likely than most to withstand the onslaught. And that is why they have reappeared in new guise in a host of unusual beliefs right up to the present day.

The Nephilim are now associated with a range of New Age and occult beliefs about aliens from outer space, vampires, werewolves, and even the Abominable Snowman (also known as Bigfoot, Sasquatch and the Yeti). The strange story of the 'Greys' (or Grays), a sinister extraterrestrial race that is secretly breeding with humankind, is a modern version of the ancient story of the Nephilim. Although grey little beings with large heads and diminutive bodies have appeared in science-fiction literature for about one hundred years, fully blown theories and beliefs about them developed when the United States entered the space age. Beliefs about the Greys, which can be found in many variations on internet sites and in magazines on extraterrestrials, fall roughly into two categories, Christian and ufologist (people who believe in UFOs). The nominally Christian view generally believes the Greys' infiltration at all levels of society is part of a plot to bring about the destruction of the human race, completing the work that the Devil started in the Garden of Eden. It stands to reason that the Greys are known for their reptilian characteristics, since they are none other than latter-day servants of the serpent.

A leading proponent in this school of thought—though with his own variation—is David Icke,[1] a former BBC sportscaster and public spokesperson for the Green Party in Britain, turned conspiracy theorist and author. David Icke's 1999 book *The Biggest Secret* proposes that reptilian aliens from the constellation Draco are plotting a new world order and causing humanity's enslavement by breeding with humans to produce a secret crossbreed of humanoids

---

1. David Icke, *The Biggest Secret* (Isle of Wight, Ryde: Bridge of Love Publications, 1999).

who are inwardly reptilian and puppets of their outer-space masters. Imagine the absorbing game that 'Grey spotting' has become, with everyone from George W Bush to Tony Blair, not to mention quite a few Jewish public figures, like Paul Wolfowitz and George Soros, identified as secret Greys.

In fact, the whole conspiratorial agenda that David Icke outlines resembles the much-circulated nineteenth-century forgery, *The Protocols of the Elders of Zion*, which became the basis of so much anti-Semitic activity in the twentieth century, including the actions of the Nazi regime, and remains a favourite text throughout the Arab world and among extreme right-wing groups. In 2001, an Egyptian twenty-part television series dramatised *The Protocols* as if it were historical fact—evidence of the tenacity of the scurrilous belief. Its screening throughout the Arab world guaranteed its ongoing popularity, and is undoubtedly the impetus of the widespread view expressed in the Arab press that Jews were both behind the destruction of the World Trade Center and the Pentagon on 9/11 and the influence on Pope Benedict XVI who made critical remarks about Islam.[2] *The Protocols* envisage the entire world in the grip of Jews, which Icke continues to pedal in his latest claim that the Jews are behind the Coronovairus COVID-19 pandemic.[3] Meanwhile, Icke's *The Biggest Secret*, is sold on Amazon and unashamedly described as David Icke's most powerful and explosive book so far. Every man, woman and child on the planet is affected by the stunning information that Icke exposes. He reveals in documented detail, how the same interconnecting bloodlines have controlled the planet for thousands of years. How they created all the major religions and suppressed the spiritual and esoteric knowledge that will set humanity free from its mental and emotional prisons.[4]

---

2. Abraham Foxman, 'Arab/Muslim Media Allege Jewish Conspiracy Behind Pope's Comments', New York, 19 September 2006: 'Anti-Defamation League press release: Cartoons and editorials published in newspapers in the Arab/Muslim world falsely portray Jews as the manipulators behind Pope Benedict XVI and his speech at a German university on the theological need for faith and reason, according to the Anti-Defamation League (ADL).'

3. BBC News 'Coronavirus: David Icke Kicked Off Facebook', Facebook issued this statement: 'We have removed this Page for repeatedly violating our policies on harmful misinformation.' 1 May 2020 https://www.bbc.com/news/technology-52501453

4. https://www.amazon.com.au/Biggest-Secret-book-change-Worldebook/dp/B00T7SK24K/ref=sr_1_1?dchild=1&keywords=David+Icke+Biggest+Secret&qid=1600769694&sr=8-1

Like the Protocols of the Elders of Zion, Icke's evil Brotherhood started in the Middle East over two thousand years ago, with the aim of controlling the whole world, starting with religion—the original poisoned chalice, according to Icke.

> In summary, a race of interbreeding bloodlines, a race within a race in fact, were centred in the Middle and Near East in the ancient world and, over the thousands of years since, have expanded their power across the globe. A crucial aspect of this has been to create a network of mystery schools and secret societies to covertly introduce their Agenda while, at the same time, creating institutions like religions to mentally and emotionally imprison the masses and set them at war with each other. The hierarchy of this tribe of bloodlines is not exclusively male and some of its key positions are held by women.[5]

The other type of belief in the Greys, promoted by ufologists, is every bit as apocalyptic as the Christian version, but it imagines that human beings will actually benefit from the implantation of extraterrestrial intelligence. After a series of painful hybridisation experiments at the hands of the aliens, human beings eventually will transcend their earthly existence in a newly evolved form.

If all this sounds too much like the frightful world of *The X- Files*, there is a bright side to the Nephilim. The union of 'the sons of God with the daughters of men' has not always produced such unsavoury visions, but in fact has seemed an appealing proposition to a fair few readers of the Bible. The prospect of a super-race, possessing the best qualities of angel and human, has prompted some romantic and nostalgic reveries, even recalling Plato's account, circa 370 BC, of the idyllic and impressive achievements of the lost civilisation of Atlantis. Indeed, Plato's account of Atlantean civilisation bears remarkable similarities to the Biblical account of the Nephilim, and begins with the god Poseidon siring five pairs of male twins with mortal women, and producing the giant race of Titans, from which Atlas emerges as their ruler.

Although the inhabitants of the island of Atlantis suffered the same fate as the first humans in the Bible, who drowned in the flood,

---

5. David Icke, *The Biggest Secret* (1999).

Plato's description of the Atlanteans differs in one significant way. Atlanteans were said to have been an advanced civilisation with peaceful ways, which is the opposite of the defective and violent human society of the first generations of the Bible. Atlantis remained a shimmering ideal in the minds of many spiritual dreamers, on a par with the Garden of Eden before the fall of Adam and Eve.

The belief that the lost race could be found or somehow revived has gained ground and exerted more-than-expected influence at certain times, such as in the late nineteenth century and again in the 1930s. During the *fin de siècle*, when scientific discoveries were placing man on the brink of a new era, there were those for whom the medical breakthroughs which promised cures for disease were not nearly as vital as the spiritual transformation which was about to engulf the Church and cure mankind of superstitious beliefs in religion. Predictably, these 'freethinkers', as they were called, believed that the advent of the new spiritual age was being hobbled by the established Churches in league with a vague assemblage of 'powers that be'. Like my dapper gentleman visitor with the top-secret documents, who believed that his information about the antediluvian races was being deliberately suppressed by powerful governments, so the proponents of the lost race were equally convinced that the Churches were too self-interested to accept what they knew to be the true origins of humanity.

One of the most colourful personalities of the late nineteenth century to advocate a lost race was the Russian-born founder of the spiritual movement known as Theosophy, Helena Petrovna Blavatsky (1831–91). A severe critic of conventional religion, particularly Christianity, Madame Blavatsky—as she liked to be known— complained that it had become littered with rival Churches of every stripe, each one more interested in filling its coffers and expanding its real estate holdings than in producing enlightened souls. In the words of a Great Adept (which is one way she referred to her clairvoyant informants), it was religion which was to blame for mankind's evil:

> I will point out the greatest, the chief cause of nearly two-thirds of the evils that pursue humanity ever since that cause became a power. It is religion under whatever form and in whatsoever nation. It is the sacerdotal caste, the priesthood and the churches . . . Ignorance created Gods and cunning took advantage of the opportunity . . . It is belief in God

and Gods that makes two-thirds of humanity the slaves of a handful of those who deceive them under the false pretense of saving them.[6]

Unlike conventional religion's preoccupation with salvation, the aim of Theosophy was edification. Its method was to invite the aspirant on a long and winding road through esoteric realms of knowledge and experience in search of Ultimate Truth. These were provided in large part by Blavatsky's prolific writings and the discussions and seances held in Theosophical Lodges, which were established early on in New York, London, Germany and Sydney.

> Meanwhile, everyone can sit near that well—the name of which is KNOWLEDGE—and gaze into its depths in the hope of seeing Truth's fair image reflected.[7]

This was spirituality modelled on the then-new public education movement, which encouraged working men and women to attend lectures on every kind of exotic topic with the aim of broadening their knowledge of the world. The difference in the case of Theosophy was that the subject matter was often decidedly esoteric. Compared to what Blavatsky labelled the materialistic concerns of the Churches, her own brand of spirituality was based on secret messages from enlightened beings, variously called the Mahatmas, or the Ascended Masters of the Great White Brotherhood, who communicated to her an eclectic blend of paranormal phenomena, Hinduism, Buddhism, ancient Egyptian and scientific discoveries, as well as occult knowledge. The latter was actually gleaned from a variety of sources, including the Rosicrucians and the Kabbalah, probably through the popular writings on occult magic by Eliphas Levi (born Alphonse Louis Constant 1810–1875). Blavatsky, who has been accused of plagiarising most of her work, nonetheless aimed to uncover a universal wisdom, Theosophy, which comprised the core truths of all religions. She also held that the spiritual insights of antiquity, if properly understood, were totally scientific.

---

6. Foreword, reprinted 1933 in United Lodge of Theosophists (ULT) Pamphlet 27, 'The Fall of Ideals by H P Blavatsky'.
7. Pamphlet 27, 'The Fall of Ideals by H P Blavatsky'.

But like the Churches, who regarded her as a bizarre curiosity at best and an agent of the Devil at worst, the scientists of her day were equally stand-offish, as they would not admit the truths so evident to the Theosophist. It must be said: Madame Blavatsky did little to endear them to her cause. With her grandiose claims and intemperate language aimed at both the Churches and scientists, she placed herself on the outer margins of the religious and scientific establishments. Yet with characteristic irony, she looked forward to the day in the near future when scientific and spiritual knowledge would be acknowledged as one unified revelation that would explain the true nature of man, including his racial heritage. Until then, Blavatsky would have to be content with disseminating her ideas through her magazine *Lucifer*.

Although she was a voracious reader and liked to cite current scientific discoveries, Madame Blavatksy was singularly unimpressed with one scientific theory making waves in her own time, Charles Darwin's theory of human evolution. She believed Darwin had it backwards. Man had not evolved from a lower to higher life form, but

Charles Darwin's theory of human evolution was declared wrong by Helena Petrovna Blavatsky, the founder of the Theosophists. She believed Darwin had it backwards and that humans had actually devolved from a higher race and not evolved from a lower life form.

had devolved from a higher state of being—that is, from a higher race to a lower one—and only later started again to ascend to a higher plane. As she explained in her book *Isis Unveiled*, the Darwinian theory of evolution took place all in the first six chapters of Genesis, where the first man referred to in the first chapter is materially different from the Adam who appears in the following verses, the former being created bi-sexed ('male and female') in the image of God, while the latter was fashioned merely from dust and was made exclusively male (and woman was created from his rib).[8] What Biblical scholars took to be two different literary versions of the one story of human creation, the founder of Theosophy construed as evidence of two distinct races. After noting this, Blavatsky turned to the Biblical mention of the Nephilim, the race of giants, whose existence she believed was corroborated by contemporary archaeological finds reported in the *Kansas City Times* of her day. The prehistoric remains of giant bones and large conical burial mounds in western Missouri and several other southern states, which Joseph Smith and others in his day took to be evidence of the Lost Tribes of Israel, were scientific proof to Blavatsky that the union of the sons of God and the daughters of men, mentioned in Genesis and described at length in the extra-biblical Book of Enoch, did indeed occur and produced a race of giants.[9]

When it came to the evolution of the human race Blavatsky had it all figured out. In her second major book, *The Secret Doctrine*, where she identified seven 'root races' of mankind, three of which had ended beneath the ocean, she claimed to have received her information from 'the Mahatmas' (Ascended Masters) who revealed to her a lost work from Atlantis, the *Stanzas of Dzyan*. In her scheme of the descending order of creation, she located the first race, which existed on the astral plane, at the North Pole. The second root race lived on the continent of Hyperborea, in the Arctic, before it sank beneath the ocean. The third root race consisted of 4-metre-tall hermaphrodites, who were highly evolved spiritually. They were from the lost continent of Lemuria, a land that had been proposed by the geologist William Bandford (1832–1905) in 1860. He located it off the east coast of Africa, south of India and north-east of Australia, but,

---

8. Helena Petrovna Blavatsky, *Isis Unveiled: A Master Key to the Mysteries of Ancient and Modern Science and Theology*, J.W. Bouton, New York, reprinted 1877; Los Angeles, Calif., The Theosophy Co. 1982, p. 303.

9. Blavatsky *Isis Unveiled*, p. 305. See also Cranston, *HPB*, p. 358.

after a giant cataclysm, it, too, had sunk beneath the ocean. (The name Lemuria was actually given by Philip Lutley [1829–1913], a zoologist who called this land bridge after the lemurs who were common to the surrounding lands.) According to Blavatsky, the Lemurians were sexually rampant hermaphrodites who produced the fourth root race, the Atlanteans. The Atlanteans evolved into fully human form with the guidance of adepts from Venus, but when Atlantis sank, they disappeared underwater. Miraculously, their offspring, the fifth root race, survived. Blavatsky identified these humans as the Aryan race, which she believed was on the evolutionary ascent, and would continue to evolve through to the seventh root race. She already had intimations of the sixth root race developing in North America, while the seventh root race would develop in South America.

The belief that the Americas held the promise of newly evolved races was a convenient stance for someone whose most recent adventures allegedly took her to North and South America, where she encountered the indigenous peoples who lived there. (The extent of Blavatsky's world travels is the stuff of legend and has never been confirmed.)

Helena Petrovna Blavatsky.

But Madame Blavatsky's interest in America waned, especially when her clairvoyant claims came under investigation and she was accused of fraud and charlatanism. She left for South India in 1879 to set up Theosophy's headquarters in Adyar, and it was there she wrote *The Secret Doctrine*, which was published quite a few years later, in 1888. This massive tome reflected her shift towards Hinduism and her adoption of its symbols, including the swastika, which she incorporated into Theosophy's emblem. In 1884, amidst further scandal about her clairvoyant messages, Blavatsky travelled to Germany and London. There her occult writings, her hierarchical racial theories, and her elitist claim to be receiving superhuman wisdom from the Ascended Masters, who were Aryan, were of great interest.

Madame Blavatsky's arrival in Germany coincided with an occult revival and a wave of romantic thinking that led to the world's most infamous experiment in spiritualised notions of race, the German search for its Aryan roots. The idea of a pure Aryan race, which had since been corrupted through breeding with the Slavic and Jewish races of Europe, offered Germans an illustrious version of their origins just when their history was beset by rival states and pan-nationalist Slavic and Jewish movements and the rise of socialist internationalism. Add to this industrialisation and urbanisation—that is, the rapid ascent into modernity—and you have all the ingredients for a popular backlash. Alternative lifestyles and rural communes flourished in an attempt to recapture something of 'authentic' German life. On the literary front, the Grimm brothers, earlier in the nineteenth century, had collected fairytales to build up a sense of Germany's unique enchanted past, its own *volkish* mythology. Around the same time, prominent philosophers, such as Ludwig Feuerbach (1804–72), inveighed against the 'Judaisation' of the German peoples, and Rudolph Marr, writing in a similar vein, established the Anti-Semitic League as a way of curbing freedoms of the newly emancipated Jews of Germany. It was a climate where the theory of a superior Aryan race would find ready acceptance.

But why Aryan? Why associate the original German race with a people who hailed from as far east as India? It was not only antipathy towards Jews and Judaism that fired German romantics to search far afield for their true identity; in the late nineteenth century the penning of anti-Jewish diatribes went hand in hand with radical critiques of Christianity, which, after all, was but a child of Judaism and therefore equally to be despised. These champions of Aryanism either imagined a future without religion, as did Theosophists, or simply aimed their criticism at the Church, in the fashion of Friedrich Nietzsche's denunciation of the 'meek' and 'pathetic' Christian God. Either way, they regarded the Church as superfluous to Germany's identity. Wotan, the ancient German god of war, was preferable to the Christian God of love, while the proud Teutonic past was incompatible with their ancient foe, the foreign Roman Catholic Church which had been transplanted onto German soil. The German future lay in retrieving the pre-Christian past, with its promise of racial supremacy.

The idea of the Aryan race arose partly as a result of nineteenth-century European colonialism. Pushing eastward for trade and political advantages had unexpected consequences. Travellers and adventurers, scholars of the Sanskrit language and archaeologists hunting for rare booty were greatly impressed by the artwork, architecture and advanced literary achievements of pre-Christian civilisations, notably the ancient Indo-Aryan peoples. The translation of Indian sacred texts by Friedrich Max Müller (1823–1900) put ancient epics like the Bhagavad Gita before a wider German audience for the first time. Franz Hartmann (1838–1912), an ardent German Theosophist who was also a translator of the Bhagavad Gita, the Daodejing and various Buddhist texts, published translations of Eastern sacred texts in his journal *Lotus Blossoms*. According to Nicholas Goodrick-Clarke it was the first German publication with a swastika on its cover.[10] The heroic and lusty narratives of the Vedas— the primary sacred texts of Hinduism going back to 1500 BC—were in complete contrast to the Protestant Christian ethos, while the high level of philosophical argument was also evidence of pre- Christian and pre-Roman sophistication.

The discovery of the Indo-Aryan civilisation, considered the oldest literate culture in the world, appealed to the Germans' search for their own pre-Christian heroic history. (Ludwig Van Beethoven was familiar with the Bhagavad Gita, which may have come to his

---

10. Nicholas Goodrick-Clarke, *The Occult Roots of Nazism: Secret Aryan Cults and Their Influence on Nazi Ideology*, New York University Press, New York, 1985, 1992, p. 25.

awareness via Freemasonry.[11]) All that remained was for someone to turn it into a romantic saga of heroes and heroines. The man who concocted this romantic history for the German people was Guido Anton Karl List (1848–1919), a self-confessed pagan with a deep abhorrence of the Catholic Church in which he had been christened. List wrote his first romantic novel, *Carnuntum*, as a paean to the 'true' history of the German people, to whom he attributed the sack of Rome in 410 by the Vandals. Although it was a fanciful history presented as fiction, it was nonetheless influential because it caught the temper of the times and appealed to the anti-Catholic sentiments of the German nationalists. A comparable phenomenon today is *The Da Vinci Code* by Dan Brown, a novel which posits an alternative version of Christianity that is no more true than *Carnuntum* was, but contains themes that mirror current agendas, such as feminism, paganism, anti-Vatican sentiment and the occult, all of which make it hugely popular with the reading and movie-going public.

If List had had anything like the marketing machine that Dan Brown did he could not have been more popular with the pan-German nationalists in his native Austria. He continued to gain a following, especially with his second novel, *Young Diether's Homecoming* (1894), about a young Teuton in the fifth century who was forcibly converted to Christianity, but ended up returning triumphantly to his pagan faith of sun worship. Later, as List began to show interest in the occult, he also claimed to be descended from Germany's first aristocratic class; that is, the priesthood of the pagan god Wotan. His public lectures about this ancient and largely fabricated history of ancient Germany gained the attention of government ministers and influential members of society, who gave their material and intellectual support to the List Society. Vienna's infamous anti-Semitic mayor, Karl Lueger, was among them.

By the time List died in a Berlin guesthouse in 1919, his ideas had been widely disseminated. According to Goodrick-Clarke in his book *The Occult Roots of Nazism*, List showed a considerable familiarity with Blavatsky's *Secret Doctrine*. He adopted her notion of seven root races, but Germanified the scheme so that the German

---

11. Brian S Gaona, "Through the Lens of Freemasonry: The Influence of Ancient Esoteric Thought on Beethoven's Late Works" https://www.semanticscholar. org/paper/Through-the-Lens-of-Freemasonry%3A-The-Influence-of-Gaona/ fd86d5aa3e4e9fdfdae644fe5b27548e8cc616fd.

peoples emerge as the Aryans. The Ario-German race was led by an exalted religious elite in possession of a secret wisdom, *a gnosis*, on par with that of Theosophy's Ascended Masters. List incorporated and elaborated Blavatsky's esoterica, including her symbols of inverted triangles and swastikas, and was so successful in this that the influential Theosophist and translator Franz Hartmann commended List on his synthesis of Germanic and Hindu doctrines.

What of those giants of old? Surely if the Aryans were to be proven the superior race, they would have to approximate the Herculean beings of antiquity? If the German adoption of Aryan ancestry was contrived, the connection to Atlantis was nothing short of bizarre. Now it was List's turn to assign religious significance to huge stones in his midst, as Blavatsky had done before him in *The Secret Doctrine*. List believed that the so-called rocking stones of lower Austria were the remains of a huge Atlantean island in the midst of the European continent. This was but another sign of many that the pedigree of the German people was the Aryan link to the supreme race of giants born of Atlantis. As Goodrick-Clarke's study makes clear, this mythology, which was elaborated by many others, had such wide appeal because it legitimised a variety of positions vital to pan- Germanism, including anti-Christian, anti-Jewish and anti-Slavic attitudes. But it offered so much more: a vision of a future when Aryans would be restored to their rightful place as the superior race among the nations. Under the zealous pen of Jörg Lanz von Liebenfels (1874–1954), the lost Aryan race assumed a scientific basis, drawing on the disciplines of palaeontology, zoology, anthropology and even electronics and radiology. Promoting an idea that would become a reality under the Nazi regime's Lebensraum program, Liebenfels advocated the segregation of 'brood mothers' fertilised by pure-bred Aryan studs, whose progeny would possess the very powers, such as telepathy and omniscience, that made Aryans closest to the gods. And if this mixture of science and myth were not contradiction enough, the ideologues of Aryanism deftly interwove the popular interest in spiritualism and the occult in their concoction of a belief system that drew on Rosicrucian, Kabbalistic and Gnostic texts as well as pagan notions of sex magic. In short, Aryanism offered a thoroughly novel, pre- and post-Christian pseudoscientific and futuristic belief system that was both utterly new yet hearkened to the deep dark past of German glory, when an imaginary Aryan race fought its enemies and won.

It was now up to its champions to demonstrate to the world that it would do so again. By the time Adolf Hitler wrote the two volumes of *Mein Kampf* in 1925 and 1926, setting out his program for Aryan domination, he was one among many racists vying for popular attention. Another who would become important to the Nazi movement was Jakob Wilhelm Hauer (1881–1962), who established the German Faith Movement based on his belief that there was a fundamental clash between two faith worlds, the Near Eastern Semitic and the Indo-Germanic, the former always ugly, the latter beautiful.[12] Hauer, a student of Buddhist and Hindu writings who had spent time in India as a Christian missionary, came to dislike all things Jewish and 'Jewish-Christian' for their moral emphasis. He soon left the Church and Christianity, which he believed was a product of the Jewish religion and therefore foreign to the unique German soul. He preferred what he believed were the amoral teachings of the *Bhagavad Gita* (in the same way that Nietzsche promoted Zarathustra as 'beyond good and evil').

Hauer grafted this new-found amorality onto a faith which he imagined as the will and experience of the soul of the German people. This 'natural religion' he compared to the Christian religious confessions and dogma which Hauer believed ruined faith. What was the purpose or aim of this German faith? None other than the glorification of the holy German people. He described the self-glorifying movement in his 1934 book *Deutsche Gottschau: Grundzuge eines Deutschen Glaubens*: 'Within us presses the power of new emotion, creative life from the holy depth of our *Volk* [people], from which all great things emerged on German soil.' As Karla Poewe's brilliant study *New Religions and the Nazis* reveals, Hauer justified his strong emphasis on the mystical and radically anti-intellectual experiences of faith by citing the iconoclastic writings of the medieval German mystic Meister Eckhart, which had been deemed heretical by the Church. Eckhart's status as a heretic would only have recommended him to Hauer, who was destined to become a key player in the Nazi SS propaganda machine. He maintained close relationships with prominent Nazis, including Heinrich Himmler, Werner Best and Alfred Rosenberg, while his founding of the United German Faith Community was orchestrated with the approval and

---

12. Karla Poewe, *New Religions and the Nazis*, Routledge, London, 2006, p. 94.

endorsement of Rudolf Hess and the two SS-Reichsführers, Heinrich Himmler and Reinhard Heydrich.

In 1932, Swiss psychologist Carl Gustav Jung, who had a significant interest in Eastern religions, joined with Hauer in leading seminars on comparative religion. Another fellow traveller was the Jewish philosopher Martin Buber, whose interest in Hasidic Judaism as a peasant mystical movement temporarily found common cause with Hauer's focus on the spirituality of the *Volk*. As Karla Poewe has detailed in her extensive study of Hauer's personal papers, he blended his Eastern predilections with a concept of merging with the Godhead, a concept borrowed from his friend and famed scholar of phenomenology, Rudolf Otto, and added to that a heroic German destiny which was awaiting fulfilment. In an essay in which Hauer argued for the reality of the German Faith, published in his journal of that name in 1934, he declared, 'It is in the Germanic *Volk* that the Indo-Germanic Faith took on exemplary form in the West'.[13] Hauer's eclectic version of this Faith promoted an Aryanised Christ and the worship of a Mother Goddess (already made popular by the Swiss anthropologist Johann Bachofen—see chapter 6), while he disparaged sin, guilt and repentance as essentially artificially engendered complexes. He wished to supplant them with feelings of union, blessedness and holiness, which he regarded as the basic religious feelings. Shortly after the seminars, the German Faith Movement was officially adopted by the ascendant Nazi Party as the official religion of Germany. Buber broke ties with Hauer and left for Palestine, where he would be one of the champions of a mixed Jewish and Arab state. The legacy of Aryan dreaming under the Nazi regime of 1933–45 would leave Europe drenched in blood and littered with concentration camps and crematoria built expressly to dispatch millions of Jews and other 'undesirable' peoples, such as the gypsies, the handicapped and homosexuals, to their deaths. It was the Nazis' fervent hope that only the Aryans would be left standing.

---

13. Karla Poewe, *ibid.* p. 72.

Chapter 8
The Allure of the East

Peter Sculthorpe had about a dozen statuettes of the Buddha and Indian divinities in his sitting room. The Australian composer confessed to me that although he had experienced monastic Zen Buddhism in Japan and later had close contact with Hinduism in Bali, he was neither Buddhist nor Hindu. Yet he did find the figurines alluring. Another major Australian composer, Ross Edwards, composed music that reflected his profound affinity with Zen Buddhism and its contemplative approach to nature. Other Eastern motifs have informed his music including Tibetan Buddhist visualisations. It is even known that the West's greatest classical composer, Ludwig Van Beethoven, was so impressed with the Bhagavad Gita, that he quoted it in his diary, and aspired to the chief attribute of Brahma: 'Free from all passion and desire, that is the Mighty One' and adopted its definition of 'blessed' man, 'who having subdued all his passions performeth with his active faculties all the functions of life.'[1]

Looking around the cultural scene today it would be hard to deny that the East has had a profound impact on the spiritual and cultural values of the West. It is fashionable to practise Buddhist meditation, psychologists quote the Dalai Lama, a yoga mat is a must-have accessory, karma is *in* while heaven is *out*, and Eastern art is second only to Aboriginal dot paintings as a major collector's item. How did this happen? It cannot be the result of Asian migration to the West because Western interest in the cultures of the East long predates the open-door immigration policies of our recent past. In fact, Western doors were firmly closed to Asians at the very time that

---

1. Quoted by Maynard Solomon, Late Beethoven, Music, Thought, Imagination, Berkeley, Univ of California Press, 2003: 170.

Europeans were gripped with fascination for their cultures, languages and religions. It is quite possible that this was no coincidence, since part of the attraction to the East was its remoteness, its utterly alien qualities and its pre-industrial way of life, a romantic view which did not necessarily encourage bringing Asians to the 'corrupting' West. It has also been argued by the American Tibetan scholar Orville Schell that the desire to enter certain Eastern countries was particularly intense when they were 'off limits' to Westerners.[2]

The allure of the East is not a recent phenomenon, and although the last two hundred years has seen a profound interest by Westerners, it can be traced back at least to the Greco-Roman period. Alexander's conquest of Persia and his march further eastward to what is now Pakistan (in 327 BC), known in antiquity as Gandhara, had culturally significant consequences, perhaps more for the West than for the East. In the latter part of the second century BC, the Greek King Manandros, who ruled western and northern India, converted to Buddhism. He has come down to us as 'King Milinda', immortalised in the book (in the Pali language) *The Questions of King Milinda*. In fact, recently carbon-dated Buddhist scrolls, the earliest known extant manuscripts of Buddhism, which came out of that region, show that the contact between Alexander's men and the resident Buddhist culture had little impact on the latter, except to introduce a Greek style of religious statuary. According to the Australian scholar Mark Alon, who is part of an international team translating the scrolls, the Buddhist texts, on the other hand, show no sign of Greek influences.[3]

With the 2001 destruction of the Bamiyan Buddhas in Afghanistan by the Muslim Taliban, even these colossal reminders of Hellenistic aesthetic influence are now gone. However, the traces of Eastern impact on the West still remain a subject of speculation. Just how many Eastern ideas flowed westward in antiquity involves more conjecture than hard evidence, but many scholars have suggested that such influence can be detected in the similarities between the philosophy of Mahayana Buddhism and Neoplatonic thought framed by the philosopher Plotinus (204–270 AD), while other scholars have argued that the ascetic practices of the Jewish Essenes and the beliefs

2.   Orville Schell, *Virtual Tibet* (New York: Henry Holt and Company, 2000).
3.   'The Ark', ABC Radio National, 26 March 2006.

of later Christian Gnostics were influenced by Buddhism.[4] The most explicit account of a Western sage travelling to India and having conversations with Brahmans and Buddhists, and adopting some of their beliefs and practices, can be found in the life of Apollonius of Tyana.[5] Born around 4 BC in Cappadocia, his travels and teachings were recorded by his disciple Damis and recounted by Philostratus in a biography completed in 217 AD (see chapter 5). Apollonius accepted the Eastern doctrine of reincarnation or rebirth, a belief that may also have migrated from Buddhism and Hinduism into Jewish mysticism and among the Karaites, a Jewish sect dating back to the eighth century who believed in the transmigration of souls. Further cultural influences would be carried west via the Silk Road trade route, that great conduit of technological innovation (especially from the Chinese), material goods, books and, possibly, religious beliefs.

But that early contact with the East was eclipsed by the subsequent history of European peoples, who vied with each other for military, religious and cultural supremacy. As the Church extended its missionary reach to claim the peoples of the European continent, converting the Goths and Visigoths in the German lands from the sixth century, fighting back the Swedish Vikings in Poland up to the tenth century and the Islamic rulers in Spain and Byzantium up to the fifteenth century, Europe became the heart and soul of Christendom. Yet even this would not guarantee peace as Catholics and Protestants were preoccupied with dividing up Christendom among themselves and building up their respective nations, erecting churches and monasteries and establishing universities. Although the East had certain commodities, such as spices, silks and precious jewels, which were sought after by Europeans, it might as well have been on the other side of the moon as far as cultural ideals went. Seen as backwaters of barbarian practices, impoverished communities and pagan religion, India and China were places with little to attract the European mind. Apart from a small number of Christian missionaries and scholars who did show an interest in the remote East, it was a place to be feared.

---

4. E Benz in *Indische Einfluss auf die Fruhchristliche Theologie* (Wiesbaden, 1951), quoted in Hajime Nakamura, *Encyclopaedia of Philosophy*, Volume I, 255; J Hastings, *Encyclopaedia of Religion and Ethics*, 1908–27, Volume V, 401; Volume XII, 318–19.
5. See Alan Baker, *The Wizard: A Secret History* (London: Random House, 2003).

Leaping forward to the late eighteenth and nineteenth centuries, a sea change in attitude could be detected throughout Europe as the French, British and German empires competed for richer lands and resources further afield and sent their colonial officers to watch over their newly acquired territories and native populations in North Africa, and the Middle and Far East. The Orientalists were born. Despite the fact that the word 'Orientalist' has acquired a jaundiced pallor due to the tendentious writings of the late Edward Said and his disciples, it was not always associated with oppressive colonial aims. In its origins, Orientalism was not a conspiracy to exploit the natives, but quite the opposite: it was a dedicated and difficult undertaking by the British initially, who believed that in order to rule India and evolve an infrastructure that was suited to Indian culture, it was important to know Indian languages, customs and religion. As Edmund Burke (1729–97) had announced to the English Parliament, Britain 'must secure prosperity of the Indian before seeking an ounce for himself', an unusually charitable approach that the historian, social reformer and parliamentarian Thomas Macaulay (1800–59) would later point out produced 'the strangest of all political anomalies'.[6]

Nonetheless, there were men full of idealism and ardour for Indian culture, who were eager to engage it with linguistic facility and read its texts with sympathetic understanding. One was the surgeon John Gilchrist, whose initial efforts to learn Hindustani (Urdu) so that he could converse with his patients ignited a passion for the language and culture. He gave up his medical career to produce publications, including a dictionary and a grammar, for new recruits of the British East India Company. Having caught the attention of the Governor-General of India, Lord Wellesley, Gilchrist was invited to help establish the college of Fort William in Calcutta in 1800, for the purpose of producing civil servants who were academically and morally qualified to serve as 'enlightened' bureaucrats. The staff at the college was a stellar cast of linguists, heading up studies in each major language and culture of India. Gilchrist was joined by HT Colebrooke, who headed the Sanskrit Department, and William Carey, who specialised in a variety of Indian vernacular languages, including the local Bangla. An expert in Chinese was also added to help with the

---

6. Quoted by Jeffrey Paine, *Father India: Western Intellectuals under the Spell of an Ancient Culture* (New York: Harper Perennial Books, 1999), 12.

translation of Buddhist texts, and a host of other specialists (up to a hundred) conducted research and published books and articles that attracted the attention of Orientalists in Europe.

Lecturers at Fort William and their graduates would add their names to the list of presidents and secretaries of the prestigious Asiatic Society, founded in 1784 by the leading Arabist Sir William Jones as the first scholarly organisation to study Asian civilisation using the new 'scientific' methods of philology, history and anthropology. The Asiatic Society in Calcutta would become the prototype for many sister organisations around the world. Although Fort William College lasted only a little more than fifty years, it has been credited with instigating a literary and cultural renaissance among the Indian elite. Perhaps sensing that this was a possible (anti- English) consequence, the college was officially closed down in 1853 due to the influence of Christian evangelicals. A parallel institution, Haileybury College, had been set up near London in 1807, drawing on resident Oxford and Cambridge Orientalists who would generate further interest in the study and appreciation of Asia.

The British East India Company, which had established both colleges, also collected a significant library of Sanskrit texts, and these attracted many scholars. Among them was the German- born Friedrich Max Müller (1823–1900), who studied Sanskrit in Leipzig and Paris in 1848 and emigrated to Britain, where he would enjoy an illustrious career at Oxford. Famous for his translation of the Brahman sacred text the Rig Veda, which, along with a commentary, took him almost ten years to produce, he would become even more so with his magnum opus, *The Sacred Books of the East*, a huge multi-volume work which became a standard source for Buddhist and Hindu studies throughout the twentieth century. For the English, these sixty-one volumes made accessible for the first time two of the world's great religions, Hinduism and Buddhism, which might have never otherwise been appreciated in all their literary and philosophical splendor. Müller, who was also deeply involved in the promotion of the new field of comparative religion, was the most outstanding Orientalist of his day. The great Irish poet WB Yeats (1865–1939), a Theosophist who became deeply interested in Hinduism, would have been familiar with Müller's translations when he attempted his own poetic rendering, with Shri Purohit Swami, of *The Ten Principal Upanishads* (1938).

Whereas Müller had been the pre-eminent translator of the sacred texts of Hinduism, Thomas William Rhys Davids (1843–1922) performed the same function for Buddhism on a smaller scale. Also trained in the Sanskrit language in Germany, Rhys Davids was posted to Ceylon (now Sri Lanka), where he became involved in archaeological work at a major Buddhist site, and collected inscriptions and manuscripts in the Pali language. When his posting came to an abrupt end, he returned to England, to work in the legal profession, but eventually gave it up to pursue his beloved Buddhist studies. He founded the Pali Text Society in 1882 and, along with his wife, Caroline Augusta Foley, who was also trained in the Pali language, began the project of translating the earliest extant texts of Buddhism. His books would become standard sources for the field, and showed an unusually sensitive understanding of the social context of the origins of Buddhism. Rhys Davids eventually took up the first Chair in Comparative Religion at the University of Manchester, and although he complained bitterly about the lack of Orientalists in Britain—Germany had twenty-two at a time when Britain had two—he was deeply appreciated by the Buddhist monks in Ceylon for making Buddhism known in the West. He would be the first of a league of eminent British Buddhologists, such as Christmas Humphries and Theodore de Bary, whose editions of Buddhist texts laid the foundations for the surge of popular interest in Buddhism in the 1960s.

It should be said that Rhys Davids and Max Müller had more than an academic interest in the cultures they studied. They didn't simply love Buddhism and Hinduism; they behaved like any passionate suitor who wished to merge with the object of his desire. Stemming from the recent research that had posited a common line of Aryan (Indo-European) languages, they both advanced arguments that there was a racial affinity between Aryan peoples and the Anglo-Saxons (see chapter 7). As a consequence, they were criticised in their day for undermining Christianity. While that was not the aim of Müller, who remained a Lutheran throughout his life and also firmly believed that Hinduism needed a 'Reformation' similar to that which had modernised Christianity, it does describe the sentiment of Rhys Davids, who believed that the British people were more racially suited to Buddhism than to Christianity. In his view, there was something fundamentally modern about Buddhism:

that interesting system of religion so nearly allied to some
of the latest speculations among ourselves, and which has
influenced so powerfully, and for so long a time, so great a
proportion of the human race—[that is] the system of religion
which we now call Buddhism.[7]

Although Buddhism has its origins in India, in Bodhgaya to
be exact, around 500 BC, and survived in Ceylon (Sielen), almost
everywhere else on the subcontinent it had virtually disappeared
by the nineteenth century. For Buddhist scholars (Buddhologists),
especially those interested in the later Mahayana tradition, China
was the place to be. Among French Orientalists Paul Pelliot (1878–
1945) reigned supreme, not only for his expertise in a wide range of
languages, but for an expedition to China in 1908 which netted him
the most important find in Buddhist scholarship: the thousands of
manuscripts in the Cave of the Thousand Buddhas in Dunhuang. In
fact, the Hungarian-born British archaeologist Sir Marc Aurel Stein
got there before Pelliot, in 1906, and purchased extraordinary cave
paintings and a large number of valuable manuscripts, including
one of the most important Mahayana texts, the Diamond Sutra. But
unlike Stein, Pelliot was a Chinese linguist and the greater scholar of
Buddhism, and thus he was able to choose carefully the manuscripts
he purchased from the custodian monks at Dunhuang. Among the
manuscripts were Buddhist texts in Sogdian, an Iranian language,
prompting him to acquire a range of cognate Persian and Turkic
languages.

Pelliot became co-editor of one of the most important journals
of Asian studies, T'oung Pao, which a hundred years later still ranks
as the leading Chinese studies journal. Earning the nickname the
'Policeman of Sinology' because of his willingness to criticise the
works of lesser linguists, Pelliot's pedantic style would never have
the broad appeal that Rhys Davids' books did, but he published
numerous articles that raised the bar very significantly in the study of
Eastern texts. More importantly, the collections of manuscripts from
Dunhuang and from other sites in China that ended up in the British
Museum, the Bibliothèque Nationale and the Musée Guimet of Paris,

---

7. From Rhys Davids Report to the Pali Text Society, cited by Lorna Dewaraja,
   'Rhys Davids, His Contribution to Pali and Buddhist Studies' (Lake House and
   Lanka Internet Services Ltd, 1996).

along with the scholars they attracted, probably saved Buddhist history from complete obscurity. In the twentieth century, when China's Communists defeated the Nationalists, under Chairman Mao Zedong almost every trace of the Buddhist heritage on mainland China was wiped out. Ironically, Buddhist studies would advance in the West, where it was free from the monastic conservatism of South-East Asia, the Hindu nationalism of India and the anti-Buddhism of Communist China.

Across the Atlantic, the East found one of its great proponents in another German Indologist. In 1938 Heinrich Zimmer (1890–1943) left Germany for England, where he taught at Oxford, before taking up a post at Columbia University in New York from 1940–43. Zimmer was clearly following in the footsteps of the late master, Max Müller, in that his interest in Hindu mythology prompted a more general appreciation of mythic narrative and the comparative study of symbols across human cultures. But unlike his countryman, who remained a believing Lutheran, Zimmer had a personal involvement in Hinduism. Not only a Sanskrit scholar and historian of Indian art, he was also a devotee of the southern Indian self-realised guru, Sri Bhagavan Ramana Maharshi (1879–1950), who numbered among his followers the influential philosopher–guru Paul Brunton and the English author Somerset Maugham, who wrote of his Indian experience in *The Razor's Edge* (1945).

Despite his career at Columbia being cut short by his early death, Zimmer would become more influential posthumously than during his life, thanks to his close friendship with Joseph Campbell. Zimmer met Campbell, who at that time was a teacher of comparative religious literature at Sarah Lawrence College in New York, when not a hint of his later legendary status was in evidence. With Campbell's interest in native American myths and Arthurian legends and Zimmer's interest in Indian and Indo-European myths they had much in common. When Zimmer died in 1943 at the age of fifty-three, Campbell produced three books from his friend's notes, *The King and the Corpse: Tales of the Soul's Conquest of Evil*, *Myths and Symbolism in Indian Art* and *Civilisation and Philosophies of India*. These works examined key motifs, such as the conflict between good and evil, and compared their treatment in Indian legend and Western sources, finding many similarities. This kind of universalism, which was the principle feature of the comparative study of religion, would be characteristic of Campbell's own research, which culminated in his first major work, *The Hero With a Thousand Faces*.[8]

As well as enabling Heinrich Zimmer's star to rise higher in America, Campbell also became one of the first American writers to make known the work of another European with a fascination for the East. Zimmer had introduced the work of his friend, the Swiss psychologist Carl Jung, to Campbell, who by that time was already interested in the psychological and artistic expressions of the myth of the Holy Grail. This was precisely the territory that Jung himself explored, along with the myths of the East and the pre-Christian West, the occult, and indigenous peoples such as the Pueblo Indians of New Mexico. The aim was to find universal psychological archetypes that constituted humanity's shared unconscious. This excited Campbell, who attended Jung's Eranos Conferences held in Switzerland. In 1953 the guest of honour at one of these conferences was the eminent Zen exponent DT Suzuki, the Buddhist scholar, who was married to American Theosophist Beatrice Erskine Lane and was enjoying a huge reputation at Columbia University in New York where he was giving a series of public lectures to packed audiences.

Both Campbell and Jung were interested in Eastern traditions, but rather than posit an East–West connection at the level of racial affinity, they spoke of a shared mytho-poetic reality or consciousness. Campbell had met and counted among his friends the Indian guru and former Theosophist, then living in California, Jiddu Krishnamurti. Jung, on the other hand, appears to have avoided contact with swamis and gurus on his visit to India in 1938, but was greatly interested in Kundalini Yoga and Tantric sex.[9] Back in Switzerland Jung experimented with these as a means of achieving psychological states associated with the four *chakras*, or locales of energy, running from the base of the spine to the top of the head. The Swiss psychologist was also fascinated by Daoism as well as Tibetan Buddhism (which is deeply influenced by Tantra), and wrote an introduction to one of its most important texts, *The Tibetan Book of the Dead*. Later, he would write an extensive foreword to *Introduction to Zen Buddhism* (1948) by DT Suzuki.

---

8. Joseph Campbell, *The Hero with a Thousand Faces*, Bollingen Series XVII (Princeton NJ: Princeton University Press, 1949).
9. Paine, *Father India: Western Intellectuals under the Spell of an Ancient Culture*.

Interest in the East is not the domain of scholars alone, although they constituted a growing presence in the humanities departments of Western universities from the early twentieth century. Before that occurred, however, one event had a momentous impact on Western awareness of Eastern culture and religion, and it took place as an adjunct to the World's Columbian Exposition held in Chicago in 1893. Sponsored by the Unitarians and Universalists of the Free Religious Association, the World's Parliament of Religions was the very first interfaith meeting in America where representatives of the world's religions gathered in their colourful splendour to inaugurate what was hoped to be a dialogue of understanding and mutual respect. It was the first time many in the West had seen Buddhist monks, Indian swamis, Sikh gurus and Islamic Sufi sheiks. The organiser of the World's Parliament of Religions was Dr Paul Carus, a scholar of religion, philosophy and science who had published several books on Buddhism, including a quaint summation of the Buddha's teachings called *The Gospel of Buddha*. (He would later collaborate on a translation of the Tao Te Ching with Suzuki, who lived and worked with him for several years.) As a result of the Parliament the first American publicly converted to Buddhism in the United States, the 41-year-old Jewish businessman Charles Strauss.

One of the stars of the event was a handsome, highly educated young Indian, Swami Vivekananda. He immediately became the favourite of Americans who were involved in the 'free thought' movement and who were interested in learning about the many ways in which an individual might achieve union with God and experience a higher state of consciousness or self- realisation. Vivekananda taught them Hinduism's most evolved philosophy, Vedanta—but with a modern twist, incorporating a strong tendency towards rational, scientific explanation. Because Vedanta asserts the Oneness of God, rather than the polytheism normally associated with Hinduism, it sounded remarkably Western. It was also compatible with an ethos of tolerance, since the emphasis on the Oneness of Existence was a perfect vehicle for promoting the brotherhood of man. This was the very purpose of the Parliament, although a century would pass before the Parliament reconvened on such a grand scale. When it did so in 1993, one of the hosts of the Parliament, the Catholic Cardinal Bernadin, recalled Swami Vivekananda, noting that his people—if not his ideas—were flourishing in the West, as Chicago had become home to a thriving Hindu community, complete with its own temples and priests.

One of the stars of the World's Parliament of Religions, held in Chicago in 1893, was Swami Vivekananda, who was popular because the version of Hinduism that he taught—Vedanta—emphasised the Oneness of God.

The impact of Vivekananda on Westerners can be glimpsed in the writings of the nineteenth-century American philosopher and psychologist William James (brother of the novelist, Henry). James's appreciation of the Vedanta philosophy was consistent with his interest in the philosophy of pragmatism as well as the psychology of religion, a field of study that James introduced to Harvard University. In a lecture on the notion of 'one in the many', or 'monism', James quoted Vivekananda's explanation of the positive effects of such an outlook:

> Where is any more misery for him who sees this Oneness in the Universe . . . this Oneness of life, Oneness of everything?
> . . . This separation between man and man, man and woman,

man and child, nation from nation, earth from moon, moon
from sun, this separation between atom and atom is the cause
really of all the misery, and the Vedanta says this separation
does not exist, it is not real. It is merely apparent, on the
surface. In the heart of things there is Unity still.[10]

When Vivekananda died at the young age of thirty-nine, Western
admirers of all backgrounds bemoaned his passing. Romain Rolland
(1866–1944), the French musicologist, essayist, novelist mystic,
and biographer of the Hindu saint, Sri Ramakrishna, who inspired
his concept of 'the oceanic feeling', likened Vivekananda's style of
speaking to the music of Beethoven and Handel:

> His words are great music, phrases in the style of Beethoven,
> stirring rhythms like the march of Handel choruses. I cannot
> touch these sayings of his scattered as they are through the
> pages of books, at thirty years' distance, without receiving
> a thrill through my body like an electric shock. And what
> shocks! What transports! Must have been produced when in
> burning words they issued from the lips of the hero![11]

Among those active in the free-thought movement who eagerly
made their way to the World's Parliament of Religions in 1893 to sit
at the feet of Eastern sages, many came from the ranks of Theosophy.
The fledgling spiritual organisation, founded by Helena Petrovna
Blavatsky and Henry Steel Olcott, was less than twenty years old
at the time of the Parliament. Although Olcott had converted
to Buddhism and Blavatsky apparently took the Buddhist vows
of a layperson, Theosophy's real aim was to foster the Universal
Brotherhood of Mankind through the study of comparative religion,
the progress of science and the belief in the fundamental unity of all
existence. Eclectic in its beliefs, Theosophy had an early connection
to India, from which it drew its beliefs in reincarnation and karma,
as well as Vedanta's monistic philosophy, which asserted the unity
underlying all diversity and all religions. Theosophy would establish

---

10. William James, 'The One in the Many', Lecture 4 in *Pragmatism: A New Name
    For Some Old Ways of Thinking*, (New York: Longman Green and Co, 1907),
    49–63.
11. As recounted by Nikhilananda, Ramakrishna-Vivekananda Center, New York, 5
    January 1953. See www.ramakrishna.org/sv

its headquarters in the south Indian town of Adyar, near Madras, in 1882, where it remains to this day.

Theosophy was influential beyond its size because it attracted a great many people in the arts, such as the Irish poet W.B. Yeats, a frequenter of the London lodge. It also attracted the upper class and the elite, such as Christmas Humphreys (1893–1983), the High Court judge who later founded the influential Buddhist Society of London, and as already noted, The Zen Buddhist scholar, DT Suzuki. Yet Theosophy's founders were an unlikely pair: Olcott, a Presbyterian, a retired Colonel with links to the founding Pilgrim Fathers of America, and Blavatsky, an occultist of Russian-German background who claimed to have travelled the world taking every sort of menial job but was privy to clairvoyant messages from the legendary Ascended Masters located in the inaccessible reaches of Tibet.

The belief in wise men who dwelled in the high mountains of Tibet had been given an ancient pedigree in the West by the first- century sage and magus, Apollonius of Tyana, who travelled to a monastery there with his companion Damis, and engaged in conversation with them. When Apollonius asked them, 'Do you know yourselves?' he was echoing the words engraved at the Temple of Apollo at Delphi ('Man, know thyself'). The head monk, Iarchas, answered, 'We know all things because we know ourselves. For there is none among us who would have been admitted to the study of philosophy had he not had that previous knowledge.' 'As what, then, do you consider yourselves?' 'As gods,' came the reply. 'And why gods?' 'Because we are good men.' Thereupon, Apollonius vowed to take the wisdom he learned there back to the Greeks, saying:

> I came to you by land and you have opened to me not only the way of the sea, but through your wisdom, the way to heaven. All these things I will bring back to the Greeks, and if I have not drunk in vain of the cup of Tantalus, I shall continue to speak with you as though you were present. Farewell, excellent philosophers.[12]

Like Apollonius, Madame Blavatsky claimed to have visited the wise men of Tibet, and was now able to communicate with them and

---

12. Alan Baker, *The Wizard*, (London: Ebury Press, 2003), 41. Also Apollonius website: Life of Apollonius of Tyana-Philostratus {220 AD} index.htm

they with her in the ether, as it were. Masters Morya (known as M) and Koot Hoomi (known as KH), also referred to as the Mahatmas, dispensed their wisdom to her on rice-paper missives, which gave her unquestioned authority on all things cosmic and divine, domestic and prosaic. When her mediumistic tricks were discovered by some insiders as well as critics, Blavatsky fled to India. Arriving in Bombay in 1879, she settled down to work on her publications while Olcott conducted his mesmeric healings, far from the prying eyes of the London tabloids. But accusations of fraud would dog Blavatsky for the rest of her days.

Theosophy had founded lodges beginning with New York in 1875, and soon after in London, Sydney, Paris, and around the world. Blavatsky and Olcott left India for London in 1884, where he petitioned the British Government on behalf of the Buddhists in Ceylon, who were being oppressed by the anti-Buddhist Hindu nationalists. She returned to India later that year to answer attacks on her by two Theosophists who alleged that her communiqués with the Ascended Masters were fraudulent. She left India for the last time, but her remaining years in France, Germany and England were marred by similar accusations. The most devastating indictment came from Richard Hodgson of the London Society for Psychical Research in 1885. Having investigated her claims, he pronounced her 'one of the most accomplished, ingenious and interesting impostors in history'. Progressively ill and depressed, she died in London in 1891, but not before installing a dynamic successor who would firmly establish Theosophy's Indian credentials.

The remarkable conversion of British socialist, feminist and atheist Annie Besant (1847–1933) to the cause of Theosophy was the most unexpected announcement she ever made to a stunned audience at the Hall of Science in Old Street, London, a gathering place for the secular-free thought movement. A friend of the Irish dramatist and Fabian socialist George Bernard Shaw, Annie's announcement was met with derision by her old mentor. Like Madame Blavatsky before her, she claimed to be receiving missives from the Ascended Masters, the Mahatmas, who summoned her to India. That they were forgeries, written in broad American slang by the head of the American division of the movement, William Quan Judge, was revealed later, but by that time Annie was set on her quest to become a champion of Theosophy

She visited Australia on a speaking tour in 1908. Annie Besant, British socialist, feminist and atheist, caused a stir when she converted to Theosophy.

and India.[13] Once in Adyar, Annie's social conscience and her political ambitions for India's independence dominated her life, whereas on matters of Theosophy she deferred to the influence of CW Leadbetter, a Theosophist and educationalist who was notorious for his predatory attentions to boys. When Leadbetter chose a poor and reputedly dimwitted Indian boy to be the mystical mascot of Theosophy, the young Jiddu Krishnamurti was undoubtedly bewildered at his good fortune and pleased to be adopted by Annie, who gave him a comfortable home and a good English education. The expectations of greatness would prove to be too oppressive, however, and Krishnamurti was embarrassed by his inadequacy and uninterested in becoming 'the World Teacher', as Annie had announced in 1925. Even at that early stage, Krishnamurti tried to distance himself from Theosophy's claim that he was the Messiah, the Returned Christ on whom the Divine Spirit had descended, but it was only in 1929 that he openly repudiated the claim. Despite remaining on cordial terms with Theosophists for most of his life, both Leadbetter and Annie had to admit that 'The Coming had gone wrong'.[14]

Annie Besant is probably better known to Indians for her social activism than for her Theosophy, although under her leadership the movement gained many followers among the Brahman classes. Her commitment to education led her to establish the Central Hindu College at Benares in 1898, and her belief in India's independence saw her found the Indian Home Rule League in 1916 and the Indian National Congress in 1917, as well as the newspaper *New India*, which advocated home rule. Indeed, it was her promotion of India's independence that attracted the young Mahatma Gandhi to Theosophy, although he would later part company with both Theosophy and Annie. Despite her strong activism, she had also established her credentials as an exponent of Hinduism when she produced an English translation of the *Bhagavad Gita* within two years of her arrival in India. Annie Besant had come a long way from her days as a devout secularist, controversial advocate of contraception, and author of *The Gospel of Atheism*.

The allure of the East would save many budding Western secularists from total atheistic oblivion, but whether they found much clarity

---

13. Paine, *Father India: Western Intellectuals under the Spell of an Ancient Culture*, 77.
14. Paine, *Father India: Western Intellectuals under the Spell of an Ancient Culture*, 87.

in their newly adopted beliefs is another matter altogether. Indeed, when Annie Besant addressed the Adyar International Convention in December 1930, she consistently stressed the need for open inquiry and free thought, and an avoidance of dogma and crystallised beliefs. Yet she warned several times that the future of the (Theosophical) Society depended on at least a portion of the members believing in the existence of the (Ascended) Masters, who remained, after all, the ultimate authority of Blavatsky's teachings. Clearly, belief in the existence of the Masters (or Mahatmas) could not be assumed and was slipping even in her day. But in encouraging this belief, she was caught on the horns of a dilemma—between devotion to her Master and her commitment to free thought. She closed her lecture to the membership with an appeal to the Buddha: 'That was the advice of the Lord Buddha, the most illuminated so far of our humanity.'

It is obvious that for Annie Besant, who spent the second half of her more than eighty years in India, the turbulent and needy nation offered a tremendous opportunity to demonstrate not just her passion for justice but impressive organisational abilities. The country's vast and complex religious terrain provided a fascinating context for her Theosophical inquiries into science and the occult, which she wrote up with her friend and colleague CW Leadbetter. In India, this unlikely pair (as odd as Blavatsky and Olcott had been before them) found refuge from Western morality, which had judged them both to be beyond the pale—Annie for her promotion of contraception and other radical causes, and Leadbetter for his unorthodox sexual preferences. British India also provided the most congenial atmosphere to promote the Theosophical dream of uniting East and West into a new synthesis that, in the language of its founder, was nothing more or less than the search for The Truth.

Hidden masters and mystics who dwelt beyond the Himalayas were not just a figment of Madame Blavatsky's imagination; they tickled the minds of many seekers after truth in the late nineteenth century. Like a previous age's obsession with discovering the fountain of youth—which the sixteenth-century Spanish explorer and governor of Puerto Rico, Juan Ponce de Leon, was determined to find in the New World—the search for higher wisdom in the modern period invariably led its intrepid seekers to the East. One such was the Russian Armenian George Ivanovitch Gurdjieff (c1877–1949) who, like Blavatsky, was something of a traveler, storyteller and embellisher of his mysterious past. He claimed to have done as many outrageous and outlandish jobs as she had, while travelling to the Middle East and Central Asia, and as far as Tibet, to find wisdom. Like her, he claimed that he had been in contact with a secret order of spiritual masters, the Brotherhood of Masters, from whom he brought back to the West an esoteric knowledge of four types of consciousness, and the method of realising what he called the Fourth Way. But fundamentally different from the rotund cigar- smoking illusionist that was Madame Blavatsky, the nimble- bodied enigma that was Gurdjieff focused on physical discipline, and to this end he elevated menial tasks to something on a par with an artful, and at times bizarre, meditation practice promising enlightened consciousness. He called his method the Work, a series of body and mind exercises which produced an ability to observe oneself in the whole cosmic scheme of things.

Gurdjieff's headquarters at Fontainebleau, near Paris, became a mecca for writers and artists, including the New Zealand-born Katherine Mansfield, the Australian-born author of *Mary Poppins*, PL Travers, and the black American author Jean Toomer, who began a renaissance of Harlem writers and spread Gurdjieff's teachings in America. The American architect Frank Lloyd Wright was deeply influenced by Gurdjieff, as was the Russian founder of the innovative dance company Ballets Russes, Sergei Diaghilev. In fact, Gurdjieff's Sacred Dances, which he had evolved from the whirling dervishes of Turkey, so fascinated Diaghilev that the impresario tried in vain to be allowed to include them as a novelty item in one of his Ballets Russes seasons. The founder and director of the School of American Ballet, Lincoln Kirstein, also made clear his indebtedness to the quixotic Armenian when he wrote in his dedication to the book *Nijinsky Dancing*, 'As in everything I do, whatever is valid springs from the person and the ideas of GI Gurdjieff.'[15]

Dancing before bourgeois audiences was a long way from the arduous journeys that some Western adventurers were prepared to undertake in seeking out the elusive quality of the East. One such adventurer was the remarkable Alexandra David-Neel (1868–1969), a Parisian woman whose youthful passion for the Orient saw her haunt the Musée Guimet, the national museum of Asian

---

15. See also Kirstein's Letter to the Editor, *Times Literary Supplement*, 27 June 1980.

art, and sit in on Sanskrit classes at the Sorbonne, for which she was not formally registered (and, it seems, ejected). She lived in the Theosophy lodge in Paris but her serious, some would say querulous, disposition probably put her at odds with some of Theosophy's fellow travellers, and she discontinued her involvement with the movement. Alexandra's greatest fortune was to marry an engineer who was stationed in Tunisia, and whose long-suffering and generous nature allowed her to fulfil a wanderlust that took her to Asia for extended periods. Her travels to the East included the Indian Himalayas, the former Buddhist Kingdom of Sikkim (now in India), China and Japan, but her real goal was Tibet. When the Chinese Qing dynasty invaded Tibet in 1910 and the 13th Dalai Lama fled to Darjeeling, she interviewed him for the magazine *Mercure de France*. His Holiness not only praised her Sanskrit but also urged her to learn Tibetan. She did so at a monastery in Sikkim, and spent the next fourteen years studying and travelling, always hoping that one day she would enter the Forbidden City of Lhasa. In 1924, after travelling two and a half years in the border regions with her trusted young monk- companion Yongden, David-Neel—disguised as a Tibetan peasant, with charcoal on her face—finally beheld the Dalai Lama's sacred palace, the Potala, fulfilling a lifelong dream. Her book *My Journey to Lhasa*, published in 1927, ended with this exclamation:

> The gods have won! The first white woman had entered Forbidden Lhasa and shown the way. May others follow and open with loving hearts the gates of the wonderland, 'for the good, for the welfare of many' as the Buddhist scriptures say.[16]

Alexandra David-Neel was taught the secrets of Tantra by Tibetan lamas, of which she wrote in many of her thirty books as well as in dispatches for Paris publications. She would become one of the most well-known Buddhologists in France, and as late as the 1960s, when she was an old woman, a long line of hippies made the pilgrimage to her home in Toulouse. Not even the Buddha attained what she did, living to 101 years old and dying sitting up in her chair. Her longevity was taken as proof that she really had learnt the secret of long life from the lamas.

---

16. Orville Schell, *Virtual Tibet: Searching for Shangri-La from the Himalayas to Hollywood* (New York: Metropolitan Books, Henry Holt and Co, 2000, 235.

In a masterful study of the lure of Tibet, the American Sinologist Orville Schell observed that the most charismatic aspect of Tibet was its secrecy and exclusivity. Despite the ban on Westerners, Tibet's proximity to India meant that the British in India would occasionally try to secretly enter the Forbidden City of Lhasa. Some of these adventurers were simply after a close look at the barbarian natives in order to spin tales which they could turn into bestsellers, like Henry Savage Landor's *In the Forbidden Land: An Account of a Journey in Tibet, Capture by the Tibetan Authorities, Imprisonment, Torture, and Ultimate Release,* which the *New York Times* described as 'a book that thrilled the world'.[17] Others were politically motivated, such as Major Francis Edward Younghusband, who, under instruction from the British Viceroy Lord Curzon, led a force into Tibet to prevent an alliance between the kingdom and Russia. Younghusband had already shown a personal interest in Tibet when he disguised himself as a Turk from Central Asia and attempted to enter Lhasa, though he was unsuccessful at that time. Leading an army in full military dress would prove to be more effective, and when he finally entered Central Tibet in 1904, the battle that ensued was, in his words, a slaughter, because the natives were passive targets, believing that the Dalai Lama would provide them with magical protection. After it was all over and Younghusband had won the race to reach the Potala, 'the citadel of Lamaism', he was at first disappointed with what he saw, and disillusioned by his meeting with the Dalai Lama. Nonetheless, on his last day there, 'the thruster', as Younghusband was nicknamed, experienced an epiphany which, as he recounted in his memoir *India and Tibet,* bathed him in the sacred beauty of the place and made him feel as if he was 'boiling [with] love for the whole world' and filled with 'an untellable joy'.[18] Later, back in London, he would contrive his own esoteric religious system based on his Tibetan experience.

Although Westerners were officially barred from entry, nothing could stop their fertile imaginations from crossing the Himalayas. In fact, Orville Schell was inspired early in life by James Hilton's novel *Lost Horizon,* published in 1933. The story follows a small group of passengers, three British and one American, whose plane crashes in

---

17. Schell, *Virtual Tibet: Searching for Shangri-La from the Himalayas to Hollywood,* 161.
18. Francis Edward Younghusband, *India and Tibet: A history of the relations which have subsisted between them from the time of Warren Hastings to 1910* (London: John Murray, 1910), 202–03.

a snowstorm while flying over the Himalayas. The survivors of the plane wreck are lost and without hope, but out of nowhere a monk appears and leads them to safety. He welcomes them to Shangri-la, a hidden paradise which Hilton clearly fashioned on the Tibetan notion of Shambhala, a mystical refuge from the pain and suffering of the world. According to legend, this mythical paradise, ruled by benevolent kings, is where the highest Buddhist teachings of enlightenment and compassion are realised on earth.

Made into a film in 1937 by the master of mystical movies, Frank Capra, *Lost Horizon* was a self-consciously philosophical critique of the material preoccupations of the West. It depicted the passengers fretting about their disrupted schedules and lost opportunities, in striking contrast to the peaceful monks who seemed to live without care yet aware of the eternal quality of every moment. The message of the film was wistfully hopeful about mankind discovering this consciousness, and it expounded the blended ideals of the Christian ethos of love and the Eastern experience of enlightenment. *Lost Horizon* was actually emblematic of the Theosophical 'project', which, above all, believed it had found the high road to Ultimate Truth, through the union of the East and West, and would pave the way for subsequent generations of Western seekers to explore that relationship further.

Chapter 9
When Religion Became Science

A purveyor of spiritual technology, SM Slavinski from Belgrade, Yugoslavia, sells his wares on the internet, promising that they are 'among the quickest and safest. They are adapted to contemporary man. They are safe. And they are cheap, quick and very, very efficient.' Indeed, he claims that 'one is able to master systems of Spiritual Technology in one day or two and to use them ever-after, through-out life.'[1]

Spiritual technology is a contradiction bordering on an oxymoron. Yet today it has become a widely adopted term, reducing the spiritual experience to a simple, repeatable, cause- and-effect technique that can be sold over the internet. To borrow from the essayist Amitav Ghosh, who was commenting on another issue, 'over the course of this century, religion has been reinvented as its own antithesis.'[2] Nothing could describe more succinctly what characterised the New Age which began in the nineteenth century. The once sharply defined differences between science and religion were cast overboard, and religion looked to its former enemy and rising star for the answers it sought.

The triumph of science has had a long genesis, and in the early modern period it had a surprising effect on religion and the human quest for spiritual truth. Rather than science entirely upending the world of faith it gave it an unexpected boost. More than a century ago,

---

1. Internet source: Sivorad Mihajlovic Slavinski, You Can Change Your Personality, Your Life, Your Destiny, Spiritual Technology of New Age. See spiritual-technology.com/eng/index1.php. See also http://spiritual-technology.com/
2. Amitav Ghosh, *Incendiary Circumstances* (New York: Houghton Mifflin Co, 2005), 120.

when clinical medicine was in its first bloom and science was on the upswing, some religious innovators were more than willing to jump on the science bandwagon, employ its language and ape its rational methods even as they denied some of its fundamental principles. Such contradictions were easily born in a period when new religions were eager to claim for themselves the immense benefits that science, including medicine, promised to humanity. It may seem strange that religion, which is normally concerned with the spiritual realm, strived to enter the hallowed halls of empirical science, but that is just what happened.

In the late nineteenth century, science was booming, buoyed by the newly established public education movement and the gigantic strides in economic prosperity that industrial technology itself had enabled. Like Thomas Edison's recently invented electric light bulb, science appeared to bring new light to every situation. The invention of the camera, the overseas telegraph and the phonograph all gave a tremendous boost to the spread of spiritualist religion, proving that unseen forces permeated the atmosphere and could be captured and materialised by scientific methods. *Faces of the Living Dead*, Martyn Jolly's invaluable study of the 'spirit photography' craze between the 1860s and 1930s, in which enterprising professional photographers provided family portraits with ghostly images of departed loved ones hovering overhead, sheds light on one of the more incredible ways in which technology fueled paranormal enthusiasms.[3] Regardless of how crude these double-exposed and doctored photographs appear to us today, with advocates such as the eminent crime writer Sir Arthur Conan Doyle, creator of Sherlock Holmes, it was little wonder that the owners of spirit portraits refused to believe the sceptics who unmasked the trickery used to create them.

When it came to science it seemed that there was nothing in the world—or in the world beyond—that its rational mind and observant eye could not illuminate and transform. Who could deny that medical breakthroughs, for example, had proved more adept at snatching the sick from the jaws of death than praying to God for a miracle? When, in the 1870s, the French chemist Louis Pasteur and the German physician Robert Koch separately discovered that germs

---

3. Martyn Jolly, *Faces of the Living Dead: Belief in Spirit Photography* (Melbourne: Melbourne University Press, 2006).

caused disease, scourges such as diphtheria, tuberculosis, anthrax and the bubonic plague were soon isolated and preventable. And when the Hungarian obstetrician Ignaz Semmelweis showed that the doctor's unwashed hands during delivery was one of the chief causes of mortality in women after childbirth, the prognosis for women improved by leaps and bounds. Death had been defeated! That was the hope. For quite a few decades, however, the reality was very different. Apart from the advances in hygiene, and the introduction of the stethoscope, invented in 1819 to detect sounds in the body, the average doctor's relationship to the rarefied world of medical research was patchy. For the most part, medical research, even as late as the turn of the twentieth century, had not yet elevated the doctor's art beyond a hodge-podge of brutal, painful and often lethal practices, subjecting desperate people to horrible indignities that usually worsened their condition, if not physically then psychologically. The penchant for blood-letting alone, which doctors applied to every kind of condition, from viruses and bacterial infections to headaches and cancer, achieved only one sure thing—the further weakening of the patient and their hastening to death's door. The unfortunate death of the beautiful and pregnant Princess Charlotte, daughter of the Prince Regent and Caroline of Brunswick, at the age of twenty-one was caused by her thoroughly incompetent doctors. Obituaries from the Victorian era, which often included graphic details of a patient's illness and death, are a revealing record of how such practices were deployed on royalty as much as on the poor.[4]

Nonetheless, the promises of medicine were exciting, despite its often scandalous and unregulated practice. For every Louis Pasteur, there were thousands of buffoons and charlatans in white coats plying their own rational-sounding theories and logical cures, whose origins lay with the second-century Greek physician Galen. One of the most entertaining exposés of this situation is George Bernard Shaw's brilliant play, *The Doctor's Dilemma*, written in 1911. Still a popular vintage film, *The Doctor's Dilemma* opens with a hilarious scene in which an assortment of doctors, each with his own peculiar theory of illness and cure-all, come to congratulate Dr Ridgeon Colenso on the announcement of his knighthood. According to one doctor

---

4. Nigel Starck, *Life After Death: The Art of the Obituary* (Melbourne: Melbourne University Press, 2006), 32–36.

who believes all illness is the result of 'blood poisoning', the universal cure or panacea is the surgical removal of the nuciform sac. For another it is the 'stimulation of the phagocytes'. For a third, a pound of greengage plums taken every day for lunch works a treat. Finally, for Sir Ridgeon, the cure for illness is a vaccine called opsonin. The terrible irony, at the end of the play, is that a large dose of the vaccine kills the patient.

Shaw, a devout anti-vaxxer, was a socialist campaigner with a bee in his bonnet about the unregulated and self-interested profession of medicine, which was held in high esteem as 'scientific' but nevertheless could deliver a lethal blow to the unsuspecting and unprotected patient. Not everyone, however, was fooled by the bumbling men in white coats with their morphine (discovered in 1804), hot suction cups and jars of leeches. One sceptical patient— who would, ironically, become Shaw's nemesis across the Atlantic— was a woman who had suffered miserably at the hands of doctors through miscarriages, chronic conditions and crippling back and neck pain. It's not certain what really ailed Mary Baker Eddy (1821– 1910)—an unhappy marriage seemed the half of it—but she became convinced that her doctors were only making her worse. A woman of independent spirit, she probably wanted to beat them at their own game. Whatever the motive, she promoted the belief that there was no such thing as physical illness, just the illusion of it. She was convinced that by giving it credence doctors approached illness the wrong way around. In what is undoubtedly one of the most unusual guides to the holy life, *Science and Health with Key to the Scriptures*, published in 1875, Mrs Eddy declared, 'The less we know about hygiene, the less we are predisposed to sickness.' Her magnum opus effectively turned the Bible into a tome of secret knowledge and, when properly understood, a medical manual. In her view, health was a godly condition that anyone might attain if only they thought and behaved in accordance with a new rational understanding of the Scriptures, which she provided. Jesus' teaching thus earned its most unlikely sobriquet, Christian Science.

The idea that one's religious faith is a prescription for good health and, conversely, that the sufferings of the physical body are but outward signs of an inner spiritual failure summarises the new revelation that Mary Baker Eddy wanted to impart to the world. In fact, it was her teacher and healer, Phineas Parkhurst Quimby,

Mary Baker Eddy believed there was no such thing as physical illness, just the illusion of it. Her book purported to be a key to reading the Bible as a kind of secret medical manual. Her views led to the formation of the Christian Science Church.

of Portland, Maine, who had taught her the art of 'mental healing', which he often referred to as 'Christian Science'. A modest man who had cottoned on to a rising interest in the pseudoscience of mesmerism, animal magnetism and hypnosis, Quimby had made a reasonable impact during his life, and in the latter years was assisted by his enthusiastic disciple, Mrs Eddy. Known then as Mary Patterson, after her second husband, she gave public lectures on 'P.P. Quimby's Spiritual Science healing disease', as noted in her letter to him dated 24 April 1864. When Quimby died in 1866, Mary Baker Eddy assumed both his mantle and possession of his notebooks, circulating and gradually revising them, and eventually appropriating his ideas as her own. (Her reliance on Quimby is a matter which the Church disputes, despite the fact that the publication of his notebooks shows it unequivocally.) Within ten years Mrs Eddy had developed her theories and established the Christian Science Church and the Massachusetts Metaphysical College.

Along the way she endured defections and lawsuits from people with genuine life-threatening illnesses who had been told they merely suffered from a lack of faith and mental fitness. But, determined as ever, at the age of sixty she moved to Boston and founded the mother church of Christian Science, still an impressive building and landmark. By 1892, the Church of Christ, Scientist had twenty churches, ninety societies and thirty- three teaching centres, and a journal with a circulation of ten thousand. Today, the *Christian Science Monitor* is considered one of America's most widely read quality papers. On the medical front, the Church's lobbyists in the United States have been successful in forty states and have, among other things, secured children's exemptions not only from routine immunisation procedures but also from school classes in biology and health, exposure to which is believed to make them sick.

It might seem ironic that a Church which believes Christ was the first scientist would steer its flock away from bona fide science. George Bernard Shaw had a simple explanation: Mrs Eddy's new religion was 'neither Christian nor Scientific', a view he probably formed in 1904 when the *Sunday New York Times*[5] published an exposé of Christian Science, revealing the founder's reliance on Quimby's notebooks and her less-than-accurate version of the events that led to her own revelation of Christian Science. But even Shaw noted that in his day science was a poorly understood endeavour. He lamented that there were still 'people for whom a man of science is a magician who can cure diseases, transmute metals, and enable us to live for ever'.[6] Mary Baker Eddy herself was captive to an overenthusiastic belief that among all the new scientific theories there was one true science that would emerge as unassailable and indisputable. (Today these are acknowledged as the signs of pseudoscience.) It would be a science of everything—and who was more likely to possess this knowledge than Christ himself, the first scientist? In truth, it would have been more fitting to call this new incarnation of Christ the first doctor,

---

5. *Sunday New York Times*, 10 July 1904.
6. George Bernard Shaw, *The Doctor's Dilemma: A Tragedy*, first published 1911, reprinted in 1947, (London: Constable and Co. Ltd), 70.

for Christian Science promised, above all, freedom from illness. By promoting salvation as an empirically verifiable state, whose physical signs are health, mental serenity and prosperity, Mary Baker Eddy succeeded in transforming religion into what she took for science.

Fundamentally, she believed that God is All and that to achieve union with the All was the destiny of true believers, who would thus leave behind what she called the pitiable illusion that is mortal man. She taught that 'Matter and death are mortal illusions', along with sin, pain and disease. This total denial of life in matter and its opposite, death in matter, with the only true life being that of the spirit, was the hidden meaning of the Biblical passages. She claimed that her own miraculous cure was affected by reading the Gospel of Matthew 9:2. But her account, which the church takes as the literal truth, contradicts the published record of her attending physician, who claimed to have successfully treated her over several months after she fell and hit her head on an icy footpath in her home town of Lynn, Massachusetts. Mrs Eddy, however, claimed that the blow to her head was so severe that 'it was pronounced fatal by the physicians'. 'On the third day thereafter, I called for my Bible and opened it at Matthew 9:2. As I read, the healing Truth dawned upon my sense; and the result was that I rose, dressed myself, and ever after was in better health than I had before enjoyed.'[7]

Although Christian Science reflected the particular preoccupations of its founder, it also tapped into the huge interest in health at the time, which also produced Ellen White's Seventh Day Adventists, another home-grown American sect founded by a woman, with a focus on health. Her spiritual regimen dictated no meat, no tobacco, no alcohol, and no caffeine. One Adventist, John Harvey Kellogg, invented the breakfast cereal Cornflakes, and were it not for the fact that Ellen White refused to invest in it, and Kellogg eventually left the Church, Kellogg's Cornflakes might have made the Church into a mainstream phenomenon. In Australia today the Seventh Day Adventists have made up for Mrs White's lack of judgement with their Sanitarium brand of breakfast cereals, although it is doubtful that many people realise they are supporting the Adventists when they buy their muesli. The idea that certain foods aroused the evil passions and caused a moral and physical decline was at the heart of

---

7. Mary Baker Eddy, *Miscellaneous Writings* (Boston: First Church, 1977), 24.

this push for a dietary path to spiritual perfection. Always interested in dietary regimes and personal health, women were greatly attracted to Christian Science and the Seventh Day Adventist Church.

The intense interest in health was but one direction in which spirituality was moving; another was towards death. The upsurge in spiritualism was occasioned by the American Civil War, and later in the early twentieth century by the First World War, in which a generation of young men was killed, leaving behind loved ones who yearned to communicate with them and learn whether they were free of pain and in heaven. Seances conducted by clairvoyants who claimed to be able to conjure the spirits of the dead and bring messages from 'the other side' were in high demand. Science and technology entered the picture thanks to the invention of photography in the 1840s; images of the Civil War with its masses of dead soldiers, had a shattering effect on all who saw them. The fact that photographs carried images of the dead gave them an eerie quality that aroused tremendous feelings, and in the hands of certain 'spirit photographers' could be transformed into proof of the spirit world.

Around the same time that Mary Baker Eddy's Church of Christ, Scientist was making headway, Helena Blavatsky was making her mark on the spiritual scene of the north-east of the United States, publishing in the Boston based *The Spiritual Scientist.* Theosophy, which she founded with Colonel Henry Olcott in 1875, was a kind of 'spiritual science' representing a higher stage in the evolution of mankind. They believed that religion's mass of superstitions and exclusive claim to salvation would be replaced by spiritual science, of the sort that Giordano Bruno, the Renaissance martyr, had advanced in his own writings. Blavatsky, who believed she was the reincarnation of Bruno (which is why Theosophy's radio station in Sydney, Australia, was named 2GB), identified with his alienation and tragic fate, believing that Ultimate Truth was 'unpalatable to most men'.[8] Like Bruno, she was destined to be misunderstood and even pilloried by her contemporaries, but for different reasons. While Bruno was persecuted by the Church for eroding its doctrine with his scientific notions, Blavatsky was ridiculed for undermining her claims to scientific knowledge by her insistence that she was receiving clairvoyant messages from the Ascended Masters.

---

8. Helena Petrovna Blavatsky, *Truth in Modern Life,* U.L.T. Pamphlet No 17, Theosophy Company (India) Ltd, Bombay, nd: iii.

Indeed, in an ironic twist of fate, the Society for Psychical Research in London, which was established to investigate scientifically her claims and other examples of spiritualism and clairvoyance, pronounced her a fraud. (It is ironic because at least half of the members of the Institute believed that spiritualism could be scientifically proven.) Regardless of its findings, which incidentally Theosophists have refuted in an examination of the original study,[9] Blavatsky did not believe that either science or religion on their own was equipped to pass judgement on her, but was convinced that, when properly understood, scientific discoveries and spiritual experience would reveal the Higher Truth that eluded ordinary men. It is at this point that mythology enters her thinking, however, since it was the Ascended Masters, occupying a higher plane of existence and located somewhere in the Himalayas, who had chosen her as their messenger to the human race. It is no accident that these mystical purveyors of spiritual science came from the East since the combination of scientific and spiritual thinking is characteristic of Indian Tantra, an esoteric philosophical and practical tradition that identifies the processes of the body, especially its divine energy, *Shakti*, with spiritual states and the cosmic order. Tantra also posits an astral (or subtle) body along with the physical body, where the former is considered real because it exists in the realm of pure consciousness and the latter is regarded as an illusion because it is material and subject to decay and death. These ideas, which have informed Hinduism and Buddhism for millennia, may seem contradictory to the modern Western empirical view of the world, but they offered an ancient precedent for the modern interest in mysterious mental powers, such as clairvoyance and hypnotism, which can effect change in the physical world.

When Madame Blavatsky relocated Theosophy's headquarters to Adyar, India in 1882, she could not have chosen a more congenial setting to pursue her ideal of spiritual science. However, it was also due to this greater identification with India that Theosophy would lose some of its early adherents. Major differences arose when Blavatsky's successor, Annie Besant, together with her friend and fellow Theosophist, CW Leadbetter, selected a young Indian boy,

---

9. Theosophy's refutation of the Society for Psychical Research's findings can be found in Vernon Harrison, *H.P. Blavatsky and the SPR: An Examination of the Hodgson Report* (Pasadena, California: Theosophical University Press, 1977).

Jiddu Krishnamurti, as the reborn Christ and World Teacher of the movement. As previously noted, Jiddu, who had been living at the Theosophical headquarters with his father and younger brother since 1909, turned out to be a most unwilling candidate, and eventually refused to fulfil his role. By then, around 1925, many early Theosophists had left or died. Among those who had been firmly against the Indian messiah was Austrian philosopher, Rudolf Steiner (1861–1925), who decided to promote his own more Christian version of the theosophical ideal.

In 1912, Steiner established Anthroposophy with its headquarters in the village of Dornach, near Basle, Switzerland. Steiner, whose most well-known legacy today is the educational curriculum promoted by Steiner Schools (originally Waldorf Schools), had a penchant for referring to his teaching as 'spiritual science'. Nevertheless, this ardent follower of Friedrich Nietzsche devoted much of his theosophical writing to the seven stages of the soul's transmigration, or rebirth, which he described in such detail that one would have to assume that he had experienced these rebirths himself and had taken a notepad along with him. Like Blavatsky before him, who mixed clairvoyant messages from the Mahatmas with reports of the latest palaeontological finds in the Arizona desert, Steiner saw no conflict between his extravagant imaginings and his pretensions to scientific method. He too was excited by the dynamic world of scientific exploration and believed that it was contributing to the spiritual evolution of mankind. But unlike Madame Blavatsky, who was a quixotic, larger-than-life prophetess, Steiner was characteristically the professorial exponent in lectures and essays of another reality, another consciousness, which he was convinced was there because he felt it and could describe it. He believed that divine wisdom lay within the grasp of human nature—hence the name: Anthroposophy— through exercises he devised which he claimed accelerated the process of human evolution or what he also called 'Earth evolution'. He believed the individual could 'become conscious of the presence of higher forces within him, which it is only a matter of developing *by appropriate exercises* [my emphasis]'.[10]

---

10. Rudolf Steiner, 'Prophecy: Its Nature and Meaning', Lecture, Berlin, 9 November 1911 (London: Anthroposophical Publishing Company, 1950), 22.

The belief that certain mental and physical exercises could train one to operate on a higher plane may well have had its source in Renaissance thinkers such as Giordano Bruno, who wrote extensively on methods of improving memory and was sought after in his day for this expertise (see chapter 2). But in the late nineteenth century such ideas were also fueled by the recent discoveries of Eastern yogic and Buddhist practices as well as Sufi mysticism, which had just become known in the West through Eastern texts translated by German, French and English scholars. These came together in Theosophy, in Anthroposophy (which accepted the Hindu notion of Karma) and in the idiosyncratic spiritual movement revolving around the self-styled practical mystic George Ivanovich Gurdjieff (see chapter 8). Gurdjieff's emphasis on work and practice had more in common with Steiner's kinetic approach. Gurdjieff, whose story is recounted in *Meetings With Remarkable Men*, devoted himself to teaching small groups a combination of body movements, dances, 'stop exercises' and mental training, which were intended to enhance inner power and outward calm, especially in work activities.

The new science of spirituality represented by Steiner and Gurdjieff is marked by a fascination with formulas and exercises in preference to petitions and prayers to a distant almighty God. In common with scientists who believed they could discover the laws of the universe, Rudolf Steiner believed that no occurrences in the spiritual domain occur without a meaning, and that it was necessary to discover the laws underlying them. Even his special interest in cosmology was characterised by patterns, especially multiples of seven, which he believed linked human life to the heavens. For example, the revolution of Saturn was said to have correspondences to the rhythm of a man's life. This he calculated as four periods of seven years, bringing one to the signal age of maturity, twenty-eight, which Steiner believed was close enough to the twenty-nine years it takes for Saturn to revolve around the Sun.[11]

Rudolf Steiner's spiritual curiosity extended to many areas, including prophecy. It bore no resemblance, however, to the Biblical prophecy that he would have learnt about in his youth, which warned of moral waywardness. Prophecy in his terms was more like the Promethean quest to attain the power of the gods in Greek mythology. Here is his hope, which he enunciated in a lecture on prophecy:

---

11. Steiner, 'Prophecy: Its Nature and Meaning', 24.

> Mankind is standing at the threshold of transition; certain forces hitherto concealed in darkness are becoming more and more apparent. And just as today people are familiar with intellect and with imagination, so in a Future by no means distant, a new faculty of soul will be there to meet the urge for knowledge of the super-sensible world . . .

> The dawn of this new power of soul can already be perceived . . . What matters is that impulses connected with evolution as it moves on towards the Future shall work upon and awaken slumbering powers in man. These prophesyings may or may not be accurate in every detail; what matters is that powers shall be awakened in the human being![12]

Steiner liked to borrow the language of scientific predictions and laws, such that spiritual development was a matter of external observation where the 'spiritual law in the flow of happenings and in the stream of Becoming' could be discerned, and would impart to one a deeper soul, a higher awareness and new powers, all of which would be developed through appropriate exercises. Yet mixed with Steiner's regimen of practical exercises was a whole mythology about the hierarchy of Angels, Archangels and Angeloi, with whom the serious Anthroposophist was meant to achieve union. In fact, the aim of the exercises was to eventually 'hand over your life to become the life of the Archangels', as Steiner declared in a lecture he gave in Cologne on 18 December 1913:

> In ordinary life we believe we think our thoughts. With training, you come to realize that the thoughts are thinking in you because the Angeloi are thinking in you. As you progress you get the feeling that you are taken to different regions of the world by the Archangels to get to know those regions.[13]

This is where Jesus of Nazareth comes into the picture. Just as the practitioner 'wants to enter into the higher worlds and read their secrets' so Steiner believed he had seen the secrets of the higher worlds, particularly the inner life and inner feelings of Jesus of Nazareth, who came into the world to promote not salvation, but

---

12. Steiner, 'Prophecy: Its Nature and Meaning', 32.
13. 'Fifth Gospel: An Explanation of the life of Christ Jesus', Lecture 13, Cologne, 18 December 1913, 213.

'Earth evolution'. In fact, Jesus himself had evolved through a series of reincarnations that included a former life as Zarathustra, the ancient Persian prophet. Steiner created an elaborate mythology of Jesus, asserting that there were two children named Jesus spoken of in the Gospels of Luke and Matthew (the latter being the reincarnated Jesus), and that they combined into one person at the age of twelve, when Jesus was lecturing the elders in the synagogue. Steiner's Jesus was so unorthodox that even he realised how peculiar it would sound to ordinary ears, which is why he pleaded that 'words have to be given new meaning for the most sacred affairs of humanity'.[14] Certainly, the straightforward message of the Christian Church that all men would be saved by declaring their faith in Jesus Christ as their Lord and Saviour was irrelevant in Steiner's elaborate scheme, which he described in more than 6000 lectures. Anthroposophists were taken on a fantastic imaginary trip and a grueling physical journey as they prepared themselves, in his words, 'through the science of the spirit'.[15]

The desire to recast the spiritual path into science was attempted again when the Church of Scientology was established by L Ron Hubbard (1911–86) in the 1950s. A science-fiction writer with an early interest in the occult and psychology, Hubbard wrote *Dianetics* (originally an article for the magazine *Astounding Science Fiction*), a book that remains required reading for Scientologists today. In it Hubbard advocated a method called 'clearing' to eradicate malfunction in human beings, which he believed was the result of traumatic experiences suffered during youth. These experiences left traces, or 'engrams', in the brain, preventing the person from exercising superpowers that lay hidden. The cure was to be effected by long and intense question-and-answer sessions by an 'auditor', who would ferret out and release the subject's alleged traumatic experiences. When *Dianetics* failed to produce the anticipated supermen, Hubbard expanded his original theory to incorporate a belief in reincarnation. This explained the previous failure to achieve clearing, since traumatic experiences and engrams could now be located somewhere in the infinite past lives of a person. It also prolonged the auditing process of clearing the engrams through an ever-growing series of required and expensive courses, which became redefined as a spiritual path in the Church of Scientology.

---

14. Steiner, 'Fifth Gospel: An Explanation of the life of Christ Jesus', 220.
15. Steiner, 'Fifth Gospel: An Explanation of the life of Christ Jesus', 216.

The name, which people often confuse with Christian Science, announces itself as a challenge to what is normally understood by those terms. Churches, after all, are the institutional symbol of 'the body of Christ', whose saving death on the cross sums up the belief and practice of Christians. Baptism, voluntary participation in Sunday services and, above all, faith, not science, is what most Churches require of the believer, whom God has freely graced with salvation. Imagine the alarm, therefore, when the Church of Scientology introduced a crude lie detector in the form of an E-meter (for electrometer) as the central feature of the auditing process.[16] The E-meter, which is hooked up to the recruit or member—who is referred to as 'pc' for pre-clear—generates a low-level electrical current in the subject, and supposedly detects 'the amount of resistance to a flow of electricity' in response to a series of questions on a Scientology questionnaire. The 'auditor' asking the questions would take his cue from these signs of resistance and either work on the pc to overcome theirl resistance, and/or suggest further courses that could help the person overcome what are believed to be the negative effects of past experiences and past lives. After a raid in January 1963, in which the US Food and Drug Administration seized hundreds of E-meters, declaring them to be illegal medical devices, Scientology was required to issue a disclaimer about the E-meters, stating that they are purely religious artefacts. With new improved designs of the E-meter, the latest being shiny red Mark VIII Ultra, Scientology has raised the bar on 'spiritual science'.

This is not to suggest that faith is removed from the picture. Hubbard devised an elaborate mythology about the origins of the world, mythical beings called Thetans and primordial volcanoes that rival anything he wrote in his years as a pulp science-fiction author. Scientologists clearly must take these stories on faith when they sign their one billion-year contracts with the Church. They are also expected to accept as absolute truth (or at least not contradict) the teachings of L. Ron Hubbard, who, like Rudolf Steiner, believed a new language had to be devised for his followers. Hubbard's own lexicon for the Church is studied and employed by Scientologists, who are forbidden to change anything he has written or taught. Public critics and dissenters are treated as recalcitrants and are punished and

---

16. L Ron Hubbard, *The Book Introducing the E-Meter* (Los Angeles: New Era Publications, 1966, 1968, 1983).

brought into line according to Hubbard's aggressive methods, which he outlined in his notorious Policy Letters. Scientologists regularly assert that followers have total freedom to believe or not believe its teachings. On the other hand, a large body of literature exists describing its disciplinary methods, as well as internet postings of verbatim Hubbard directives. There are also investigative reporters' press articles, studies by academics and first-hand accounts of ex-Scientologists produced over the past fifty years which show a Church that takes quick and punitive action against its vociferous critics and 'apostate' former members.[17]

In the end, to echo the words of George Bernard Shaw, what is taught is not really science, which is always open to critique and revision, and cannot be mixed with or grafted onto a highly imaginative mythological belief system. Nor is Scientology a Church in the Christian meaning of the term—that is, a place where the Gospel is freely given to strangers as well as parishioners, with no obligation to either pay a tithe or drop a few dollars in the basket after the sermon is over. On the contrary, Scientology's spiritual path, which is known as The Gradient.

The language of this new kind of religion is consistently kinetic and aspirational—all about learning how to achieve superior functioning and to develop never-before-imagined abilities. To a generation used to taking speed and other mind-altering drugs in order to accelerate performance, Scientology's promises, together with its anti-drug campaigning, is an intriguing alternative. No wonder it is so popular in Hollywood, the high-anxiety capital of the world. When Andrew Denton host of ABC TV's *Enough Rope* asked leading Scientology advocate and film star John Travolta to explain what Scientology did for him the Hollywood actor said that it was being relaxed in front of a live audience, fearless, able to perform with total confidence, and knowing that the people out there loved him (big cheer from the studio audience). It is significant that Travolta chose to talk about the career benefits of being empowered and successful. The attraction of

---

17. Janet Reitman, 'Inside Scientology', *Rolling Stone*, February 2006. But academics have also lent their support to Scientology, and their 'expert testimony' is sought for court cases. British sociologist Bryan Wilson, for example, prepared a report which Scientology regularly sends out to the media on the unreliability of evidence from former 'disgruntled' Scientologists. See, Rachael Kohn, 'The Return of Religious Sociology', *Method & Theory in the Study of Religion* 1/2 (1989): 135–159.

this and other forms of spiritual technology is precisely their promise to deliver such objectively measurable benefits to the user.

Scientology has been so financially successful in delivering rafts of courses which followers self-fund, either through cash or in exchange for work, that it has inevitably spawned imitators. This is probably one of the reasons why Scientology took the unorthodox step of trademarking its courses, which caused quite a stir in the religious world, where spiritual teachings are offered freely as God's gift, not as commercial products. The other reason for trademarking them, however, was to prevent ex-Scientologists from posting them on the internet for the non-paying but curious public. (There is nonetheless quite a significant amount of Scientology material posted on the internet under the 'fair use' clause of copyright law. See, for example, Secrets of Scientology.[18]) One controversial Australian sect, known as Kenja, founded and run by ex-Scientologist Ken Dyers (1922–2007) and his wife Jan in 1982, uses some similar 'processing' methods that promise to improve everything from business management to interpersonal relationships, as anyone will discover when they attend one of their public forums held in the hope of signing up new 'users'. However, not even the promises of Kenja could prevent Dyers from taking his own life when faced with a trial for twenty-two offences of alleged sexual assaults he committed on three underage girls.[19]

Scientology and its various offspring offer the new recruit freedom from malfunctioning through spiritual technology, a term that has become a popular label for all sorts of home-grown methods that contain a fair swathe of self-hypnosis techniques. In fact, the list of spiritual technology spruikers and users grows by the day, as our material culture becomes more enamoured of technological innovations and our expectations of personal prowess are heightened. The Synchronicity Foundation, created by Master Charles, formerly Brother Charles, has developed a trademarked range of products that deliver 'the Synchronicity Experience' utilising 'precision Holodynamic Vibrational Entrainment Technology (HVET)'. Based on the premise that meditation is good for you but, in its traditional form, is unreliable, uneven, and just plain difficult, this self-declared

---

18. On Operation Clambake, Undressing the Church of Scientology since 1996 on www.xenu.net

19. https://www.smh.com.au/national/nsw/widow-of-cult-leader-loses-case-against-nsw-police-over-his-suicide-20200605-p54zxh.html

'Modern Mystic' offers a series of soundtracks that takes the fuss out of meditation and guarantees results: 'It meditates you' by balancing the brainwaves and delivering a precision meditation experience—one that is accurate and consistent. And depending on whether you want to meditate 'lite', medium or in-depth, there are Alpha, Beta and Delta frequency ranges available. Log on to the website, open an account, and hey presto, 'a modern, holistic, spiritual lifestyle method that delivers greater balance and wholeness' can be yours. Just make sure you have your credit card handy.

Apart from the profit motive, the obvious appeal of spiritual technology is that it is not entangled in the messy world of ideas, there are no religious rituals and moral imperatives to learn, and it does not require a communal commitment or face-to-face encounters. The CDs, the downloads, the emails and the telephone calls with trained facilitators are all that is required for the ongoing synchronicity experience. Holistic bliss is the aim, but the method relies on empirical evidence of brainwave patterns. Spirituality has become a clinical exercise. As the promotional material says, 'The Synchronicity Experience is based on the current interface of modern mysticism, science and technology.' The great I AM has turned into the great iPOD.

Chapter 10
Utopian Dreams

I want the world to know that we have ninety-three Rolls-Royces because that is the only way to make any bridge. And then I can talk about truth and enlightenment, too, on the other side. Without Rolls-Royces there is no communication at all. I know my business perfectly well.[1]

They used to be called communes in the heady 1960s and 1970s, when groups of hippies flouted social conventions and religious mores to live sexually free lifestyles with multiple partners and goods in common. Their moral values were more often than not unorthodox, but their radical communism implied a spiritual striving that borrowed heavily from the revolutionary examples of Jesus, Mao and Gandhi. At least, that is what their rhetoric suggested, regardless of how little they understood their inspirational heroes. Rajneesh, the founder of Rajneeshpuram, seemed to keep his followers in thrall precisely because he took great delight in subverting conventional thinking on most issues, including wealth, sex and Christianity. Nevertheless, today the commune experience is often remembered nostalgically as the epitome of youthful idealism. Occasionally, however, an older and wiser 'survivor' of those social experiments sheepishly relates a searing tale of a utopian dream that turned into a nightmare, where open relationships took a heavy emotional toll, the collective raising of children resulted in alienation from them, and charismatic leaders trampled on basic human rights and needs. At times, leaders even encouraged serious crimes, such as murder, extortion and child abuse.

---

1. Bhagwan Sri Rajneesh, the founder of the utopian community Rajneeshpuram in Oregon, USA, quoted in *Osho International Foundation, Osho: Autobiography of a Spiritually Incorrect Mystic,* (New York: St Martin's Press, 2000).

The worst such case in modern times, was Jonestown, a commune in the jungles of Guyana, which ended in 1978 when 914 people—mostly black Americans, and 276 of them children—were forced to drink cyanide-laced grape drink while their leader, Jim Jones, an ardent white communist addicted to barbiturates, ranted that they were about 'to go to God'.

Jim Jones had founded the People's Temple, a full gospel church with a radical socialist message and an anti-racist agenda in the 1950s. But investigations into his claim that he could heal cancer in his miracle prayer sessions compelled him to move his church from Indianapolis, Indiana, to a small town outside San Francisco. Always sailing close to the wind, Jones was again investigated for a number of illegal activities, so he moved more than 900 of his followers to the jungles of Guyana with the aim of building a socialist paradise. But stories of concentration-camp conditions alarmed people back home, and US Congressman Leo Ryan visited Jonestown to make inquiries. On 18 November 1978, while waiting at the local airstrip to board a plane back to the United States, Congressman Ryan, three members of the press and about a dozen defectors were fired on by heavily armed guards of the People's Temple. Ryan and the press corps were killed, as well as one person from Jonestown, and eleven of the defectors were wounded. Jones panicked, and later that day orchestrated the deaths of all his followers, as well as his own. Jonestown had lasted barely two years.

Other communistic experiments ended just as badly, such as the mass suicide of thirty-nine members of Heaven's Gate in Rancho Santa Fe, California, in 1997, and the inferno at the Waco, Texas, compound of the Branch Davidians in 1993 which left about eighty members of the cult dead, including its leader David Koresh (aka Vernon Howell).

Even when paradise is lost, it is not always easy to admit it. A case in point is Rajneeshpuram, in Oregon, which went into meltdown in 1985. The community's spiritual leader, Bhagwan Sri Rajneesh, was on the run from the United States Federal Police for breaking immigration laws, while members of its leadership were charged and later imprisoned for attempted murder and a host of other crimes, including the fire-bombing of a county records office, and the first case of 'bioterrorism' in the United States, when salmonella bacteria was dumped in the salad bars of ten local restaurants in The Dalles,

Oregon, resulting in the poisoning of 751 people. It was a long way from the universal love Bhagwan Sri Rajneesh preached to his followers.[2]

At first, Rajneeshpuram looked like a hugely successful community built around the eccentric Indian philosopher guru who blended popular psychology, free love and eclectic spiritual instruction with a breathtaking self-indulgence in personal adornments from Rolex watches to Rolls-Royces. But like a fast- growing plant with shallow roots, its life cycle lasted only four years. In 1981 a 64,000-acre (26,000-ha) ranch was purchased, and within three years it had been transformed into a thriving community of 7000 residents with its own farm, school and post office, as well as 15,000 visitors annually. It had ambitions to take over the neighbouring town of Antelope, which was renamed Rajneesh in 1984, but the town and country residents resisted, which is what prompted the cult's criminal offensive. It did not take long, however, before it all came to a messy end. By 1985, the community's leadership had absconded with millions of dollars, leaving Baghwan Sri Rajneesh, 'the laughing guru' with his ninety-three Rolls-Royces, to face criminal charges. He was let off lightly,

---

2. Twenty years later, one legal case was still on the books, the extradition of West Australian Catherine Jane Stork to face charges of conspiracy to murder the chief federal prosecutor appointed to head a federal grand jury investigation into the commune. Stork, who had gone under several surnames, Paul/Lalor/ Elsea, had already been convicted of attempted murder of a physician, but had received a reduced sentence of less than three years through plea bargaining. Having returned to Western Australia to live with her parents, she later moved to Germany, where she became Catherine Jane Stubbs, and then married, becoming Catherine Jane Stork. She would not escape her past, however, when it was discovered in 1990 that she had been part of a conspiracy to assassinate Federal US Attorney for Oregon, Charles Turner. Extradition proceedings followed, but Germany proved a reliable protector, and the now-German citizen was not turned over to the United States authorities. Sad news of her son's terminal illness, however, compelled her to return to the United States in September 2005 to face the outstanding criminal charges, in a bid to receive a suspended sentence (the assassination plans were never carried out), so that she could visit her son in New South Wales. Expressing some remorse for her actions and regret for a life gone wrong, she was shown mercy by the judge, who imposed a lenient sentence of five years probation in Oregon, to be served after she was granted leave to visit her dying son in Australia. Stork's story, available on the internet, was most recently investigated by Richard Guilliat, 'It was a time of madness', *The Weekend Australian* magazine, 17–18 June 2006. Janet Stork was my guest on 'Bahagwan Blues', The Spirit of Things, April 26, 2009. https://www.abc.net.au/ radionational/programs/archived/spiritofthings/bhagwan-blues/3144872.

pleading ignorance of what his subordinates were up to, and returned
to India. He regrouped under a new name, Osho, living no less
lavishly than he had in Oregon. He died in 1990.

Despite the obvious failures of Rajneeshpuram, and the often-
nonsensical pronouncements of its guru, loyal former members,
sannyasins—popularly known as 'the orange people' because of
their orange clothing—still remember their sexually experimental
utopian community with fondness. Most have refused to sheet home
the blame for its collapse to their beloved guru, who, after all, gave
them the irresistible message that sexual orgasm is an experience
of bliss on par with meditation and is in fact the gateway to God-
consciousness. He gave them an ideology but also a lifestyle that was
every teenager's dream: summer camp with sex and no guilt. As he
said on *Good Morning, America*, 'Yes, I believe in free sex. I believe
that sex is everybody's birthright to share, to enjoy. It is fun. There
is nothing serious about it.'[3] There are Sannyasins who continue to
champion Osho's philosophy, living communally and still wearing
the guru's favourite colour, orange (saffron being the sacred colour
of Hindus and Buddhists) their youthful idealism and sexual identity
forever indebted to the man who embedded sex in spirituality.

For many believers, the idea of utopia is truer than its harshest
realities. But for those who are a little embarrassed by its rather
ignominious reputation, a new term for communal experiments has
become popular. The label 'intentional communities' counters the
notion that communal groups might have been founded on naïve
optimism or coercive totalitarianism, and that no matter how they
panned out in the end, they were created by design. It is a term
calculated to downplay the fact that many such communities meander
far from their original idealistic aims and promises, whether hijacked
by power-hungry leaders or simply were poorly conceived in the first
place. Indeed, history is littered with the ruins of failed utopias.

Ironically, the idea of utopia was originally nothing more than a
standard by which to measure and even criticise the shortfalls of society.
That was the intention of Thomas More, the Lord Chancellor under
Henry VIII, who first put Utopia on the map, as it were. Inspired by
Plato's *Republic*, the ancient Greek philosopher's blueprint of an ideal
democratic society, More penned a tale about a land called Utopia,

---

3. Osho International Foundation, *Osho*, 130.

which he located in the recently discovered New World. Like Plato's fictional republic, More conceived Utopia as a clever device to explore the practical ways in which society could maximise the common good and prevent the economic deprivation and resulting moral turpitude which he saw as endemic to England and much of Europe in the sixteenth century. It was an interest shared by many others, given the immediate popularity of *Utopia* when it first appeared in Latin in 1516. *Utopia* was published in Paris in 1517, in Basel in 1518, in Florence in 1519, and in subsequent Latin editions throughout the seventeenth and eighteenth centuries. There were also German, Italian, Dutch and English translations in the sixteenth century. At the very least, Utopia's popularity all over western Europe indicates how seriously questions pertaining to governance and the role of the

Thomas More, Henry VIII's Lord Chancellor, wrote Utopia about a land he located in the freshly discovered New World. The notion of a place that could be founded for the common good was immensely popular in Elizabethan England, and ever since. Courtesy National Library of Australia.

King, his Council and the Church, were being debated at the time. A century later they would preoccupy England's greatest playwright, William Shakespeare, for whom royalty is a major preoccupation, from its pomp and glory to its murderous lust for power.

In More's day, no less than in Shakespeare's, Europe was beset by religious strife, with deep social divisions between decadent aristocrats and poverty-stricken masses who, in England, were forced to move to the cities due to the vast 'enclosures' of land for sheep farming. In contrast to this rapidly changing era that also saw the beginnings of large-scale manufacturing, monopolies and impersonal markets, More imagined an island country, which was about the size of England and Wales combined, where the inhabitants were all engaged in agrarian work, held their property in common, despised money and pursued the public good. Every citizen enjoyed equal status, including women, and the great gaps between rich and poor did not exist. There were few opportunities in Utopia to lord it over another person, even in matters of the spirit, as Utopia subscribed to complete freedom of religion. More described it as 'the only commonwealth that truly deserved that name', but it was in fact a mythical vehicle for his diatribe against the social inequalities that prevailed in England and the injustice of peasant life as is evident in this excerpt:

> I would gladly hear any man compare the justice that is among them [the Utopians] with that of all other nations; among whom, may I perish, if I see anything that looks either like justice or equity; for what justice is there in this, that a nobleman, a goldsmith, a banker, or any other man, that either does nothing at all or at best is employed in things that are of no use to the public, should live in great luxury and splendour, upon what is so ill acquired, and a mean man, a carter, a smith, or a ploughman, that works harder even than the beasts themselves, and is employed in labours so necessary, that no commonwealth could hold out a year without them, can only earn so poor a livelihood, and must lead so miserable a life, that the condition of the beasts is much better than theirs? . . . Is not that government unjust or ungrateful . . .?[4]

---

4. This translation is preferable to the stilted version in the edition of *Utopia* edited by David Harris Sacks, although his introduction is excellent. See Sir Thomas More, *Utopia*, edited with an introduction by David Harris Sacks (Boston: St Martin's Press, 1999), 198–99. See also the website http://oregonstate.edu/instruct/ph1302/texts/more/utopia=II.htm

More was well known for his close friendship with the Protestant sympathiser and Dutch Christian humanist Desiderius Erasmus (1566–1636), who was an outspoken critic of his own Church's cruel dealings with heretics and its suppression of freedom of conscience in theological matters. More, on the other hand, was a loyal Catholic, perhaps more so for being in England, where the sanctity of the Church was challenged by King Henry's personal agenda. Indeed, More, who partook in monastic life in a Carthusian monastery while he was practising law, never lost his high regard for the unworldly pursuit of religious contemplation, even when a life in politics beckoned. It is little wonder, then, that More did not to accede to his sovereign's demand to be made the 'supreme head on earth' of the English Church by signing the Oath of Supremacy, and refused to swear to the Act of Succession, which put the Protestant Elizabeth next in line for the throne. By his refusal, More sealed his fate. He was sent to the Tower of London and executed on 6 July 1535, his last words on the scaffold being, 'The King's good servant, but God's First'. How ironic that during those turbulent times, when the Roman Catholic Church and its infamous Inquisition was in the firing line from More's humanist fellow travellers, it was the English King who ordered More's death so that he might establish his new Church.

More's legacy would live on, however. Utopia's central argument, that sharing goods in common is the key to social peace and prosperity, would recommend itself to many thinkers—including, most famously, Karl Marx. Not that the principle of social justice and the practice of charity had been alien to Western society, for both were important Biblical teachings. But in Utopia the controlled distribution of wealth was built into society itself.

It was a cosy vision in which the whole society was organised as if it were one family household, with all goods in common, where everyone was satisfied and consequently behaved with loyalty and love. As a result, personal vices, such as pride and greed, as well as social ills, such as corruption and theft, were eliminated, while provision for the weak and impotent was freely given.[5] Added to this, the Utopians' personal characteristics were unparalleled. They had an overwhelming love for learning, with which they filled their free time, while their pleasures consisted in the prudent cultivation and

---

5. Thomas More, *Utopia*, 198.

enjoyment of good health. Gluttony, greed and lust were nowhere to be seen. Even the Utopians' fundamental religious freedom was tempered by a wise acknowledgement that despite their various beliefs and practices, which they observed in the privacy of their homes, they all worshipped the one true God at a common Temple. Religious conflict was not just unknown but unnecessary. Utopians all wore white, which is fitting given their island resembled more a republic of angels than a society of human beings.

One need hardly point out the naïve assumptions and romantic delusions on which Utopia rested. Indeed, More's own gentle doubts about the viability of this system, including the elimination of private property, were appended in a dialogue he created between himself and his narrator, Raphael Hythloday, who was an eyewitness to this unique community in the New World. Comprising Book One, this friendly exchange also contains a discussion about the roles of the King and his counsellors in governance, which undoubtedly reflected the fact that More was at the time considering taking up a position as a member of the King's Council. When More was appointed Master of Requests in 1518 and was in regular attendance to the King, he became well aware of the harsh realities of political life both on the domestic and foreign fronts. The intrigues of the royal court would have made one thing absolutely clear: that the just society was a worthy aim but at best only a partially realisable goal. Utopia's fundamental flaw was, of course, that there is no social structure that can successfully erase evil from within the human community or human heart. More, who wore a hairshirt under his clothing as a reminder of his monastic ideals, must have known that that province was God's alone—hence the naming of the mythical island 'Utopia', which in Greek roughly translates as 'no place'.

If the human passions, which seek gratification, are a constant threat to the harmonious community, so too are they the vehicles of great hopes, and none more enduring than the search for the perfect society. Indeed, the cautionary voice that More wove into *Utopia* spoke softly compared to the enthusiastic reports of Hythloday, whose description of the social paradise he had witnessed in the New World inspired none other than Vasco de Quiroga, an educated Spaniard of noble lineage who had served his country in Africa before being sent to New Spain (Mexico) in 1531. A judge, his task was to oversee the development of Spanish interests in the New

World, and to stop the indiscriminate slavery of the native peoples which his predecessors had practised with dire results. Quiroga took with him a copy of More's *Utopia*, believing that it was a blueprint of 'a holy commonwealth that Christians ought to imitate',[6] and it inspired him to organise the indigenous peoples into economically self-sufficient communities, which he called 'republicas'. He copied much from his English mentor, including the white clothes for his people, the number of hours worked daily in the fields and the name of the dwellings, the 'familias', which echoed More's notion that each Utopian town's building block was like a family. He also taught the indigenous people a number of crafts, which they still practise today.

In great contrast to the communes of our era, which were started by an affluent generation who opted for a simpler life, Quiroga's families had the far more urgent task of rescuing the native Mexicans from slavery, starvation and the scourge of smallpox. Quiroga, who had became a priest and later a bishop, also used the opportunity to bring to the natives a religion that would replace their practice of ritual human sacrifice with the Christian message of a God who had sacrificed his only son to save the world. Together with the hymns that Quiroga wrote for them, and which they sang in their own language, the new religion would prove to be a very attractive substitute for their old one. They continued to celebrate some of their traditional festivals, which Quiroga described as colourful and merry occasions. Although he was pensive about but not intolerant of them, one wonders how much these demonstrations prompted the aging Spaniard to ponder the viability of More's religiously free Utopia. In any case, Quiroga was not discouraged from bringing Christianity to the natives.

One thing was certain: More's emphasis on health as an important source of happiness appealed to Quiroga's practical bent. He built a major hospital, Santa Fe de la Laguna, as well as a network of hospitals known as *hospitals de la concepción* (named for Santa Maria de la Concepción) so that every town in Michoacán had a place for the ill and the poor to receive food and the Holy Sacraments. Even more than the churches he built in Michoacán, the hospitals would increase and prosper throughout the region, such that by the end of

---

6. Quoted in Toby Green, *Thomas More's Magician* (London: Wiedenfeld & Nicolson, 2004), 130.

the seventeenth century there were 264 pueblo hospitals, which he had built and run with the help of the Franciscan mendicant friars.

Quiroga is remembered as a saint by the people of Santa Fe today, but he spent most of his life battling the bishops of Mexico and Guadalajara, the commercial interests, the *encomenderos*, of Michoacán, and even the Franciscan missionaries of his see, whom he thought expended too much effort building monasteries for their own benefit. Yet he had far fewer quarrels with the native peoples than the other bishops of Mexico, who had brought the full force of the Spanish Inquisition to bear on the newly converted natives. They were accused of 'judaising', a preposterous accusation on the face of it, but understandable at a time when the Spanish believed that the native peoples of the New World were remnants of the Lost Ten Tribes of Israel (see chapter 4). There was only one recorded case of a heretic accused as a judaiser in Quiroga's see, because on the whole he focused on the welfare needs of his people, the Purépecha, rather than on their theological purity.

More than one outbreak of smallpox would devastate, but not destroy, them, ensuring that Quiroga's achievements among the Purépecha were remembered long after his death, which occurred on 14 March 1565 at the age of ninety-five, when he was riding a mule on one of his regular tours of the diocese. His last will and testament was largely devoted to his beloved network of pueblo hospitals, and especially the major hospital, Santa Fe de la Laguna, with its supporting farm. But records show that he left a legacy other than hospitals. 'Tata' (Uncle), as he was known among the Purépecha, had taught them well; when a new and different style of leadership was introduced by Quiroga's Spanish successors, they tolerated it only up to a point. When they decided it was not to their liking they rose up, not in battle, but in orderly delegations, presenting petitions and requesting justice through legal processes 'with as much liberty as if they had been Spaniards' according to an incredulous priest in 1581.[7] This was a fair indication of how assimilated the native people had become to the methods of governance which Quiroga, like his mentor, Thomas More, believed were the key to a harmonious and healthy society.

---

7. Green, *Thomas More's Magician*, 276.

Looking back on Quiroga's utopian communities, the question is why did they succeed when more-recent efforts have not? Certainly, it helped that the Purépecha knew they had a lot to gain by joining Quiroga's *republicas* compared to the fate which many of their countrymen met when they fled into the jungle, only to be caught and enslaved in the silver and gold mines, worked to death by rapacious masters who did not regard them as fully human. Quiroga may have been a solemn man and also querulous, but he believed the natives of Mexico to be fully human and gifted with a peaceful outlook, which is why he preferred to ordain them as priests over the Spaniards, a decision that led to trouble with the Church hierarchy in Spain. Yet he was not a renegade, out to establish his own Church with himself as its leader. He was a loyal servant to the Holy Catholic Church, but distant enough from it to try out some new ideas. The combination of the native people's real need, the lack of a humane alternative, and their innate character, combined to make the experiment worth pursuing and, with such a long-lived and singularly devoted leader, it was bound to succeed.

Long after Quiroga, the ideals espoused in *Utopia* inspired a host of earnest imitators—especially in the nineteenth century, when New Age health farms and socialist communes competed with traditional Church-based religion for the hearts and minds of the emergent middle class. Karl Marx was certainly among the intellectual converts to the idea that private property was the root of all evil, even though More himself questioned the practical wisdom of that view. Marx also bore considerable contempt for religion, which could never be said of More. Nevertheless, communistic experiments that jettisoned traditional religion were the rage, and it is precisely because of their often-bizarre manifestations that Thomas More's *Utopia* also attracted its share of satirists. One such was the British author Samuel Butler (1835–1902), who alluded to More's theory of health-equals-happiness when he described existence on *Erewhon* (1871), another fictional community rich in the bounties of nature: 'It was a monotonous life, but it was very healthy; and one does not much mind anything when one is well.'[8] Suddenly 'wellness' seems a bit too dreary to be truly desirable.

---

8. Butler, *Erewhon*, 1871, internet, Chapter 1. https://www.britannica.com/ biography/Samuel-Butler-English-author-1835-1902#ref47597

Butler might have been describing his own experience as a colonial 'down under'. A son of an Anglican vicar, he had high hopes of following in his father's footsteps and was educated at Cambridge, getting a First in Classics. Afterward, during a period of teaching in a poor parish school, Butler became disillusioned with the Church's teachings and did not seek ordination, much to his father's disappointment. Perhaps to put some distance between himself and the shame of breaking a long and distinguished family tradition (his grandfather had been a bishop), Butler went to New Zealand to make his fortune in sheep farming. It was during his five-year sojourn there, from 1859 to 1864, that he was inspired to write *Erewhon* (a word play on 'nowhere'). His description of the beautiful but often gloomy island is an unmistakable allusion to what the Maoris call the Land of the Long White Cloud.

Across the Tasman, one writer who was inspired to pursue a utopian dream with gusto was William Lane (1861–1917). Born in Bristol, Lane travelled to the U.S. and Canada before arriving in Brisbane, where he became a prolific journalist writing under various pseudonyms in a variety of areas. But it was labour issues and trade unions that were his passion in columns for the Brisbane *Courier* between 1883 and 1885, and the *Observer*. He began the magazine *Boomerang* in 1887 to allow himself free rein to promote land settlement and land reform as a means of solving the social problems resulting from unemployment. After three years, he sold *Boomerang* and became the editor of *Worker*, and from this would become the prime mover of the organised labour movement. Always restless, Lane was soon on to his next venture, writing the novel *The Working Man's Paradise* (1892), published under the pseudonym of John Miller, which laid out his design of a socialist utopia. Unlike Thomas More, who was content to write about it, Lane began to investigate the possibilities of realising his dream.

Like so many other architects of utopian communities, William Lane was convinced that the first thing to do was to remove himself from the corrupt world and start afresh—the pristine forests and plains of Paraguay would do just fine. Having collected funds and purchased land in the name of The New Australia Co-operative Settlement Association, 220 pioneers sailed from Sydney on the *Royal Tar* in July 1893 and arrived in New Australia, almost 2000 kilometres up the River La Plata, in October of that year. Clearing

Australian journalist William Lane promoted the utopian ideal to such an extent that he founded a colony—New Australia—in Paraguay to start a workers' paradise.

and planting were the immediate tasks, but soon Lane's strict ban on alcohol and consorting with local 'coloured' women created conflict in the group. When another contingent of settlers arrived, the problems grew worse. Even his admirers admitted that Lane was an authoritarian who burned with the purity of his ideas but exercised them with inflexibility, little tact and no consideration for people's feelings. When it appeared to them that their leader was pursuing not just a socialist dream but a racially pure Anglo-Saxon one, there was a rebellion, and the dissenters sought the assistance of a British diplomat at the British Legation in Buenos Aires, Mansfeldt de Cardonnel Findlay. On meeting Lane, Findlay filed a report:

> Mr Lane has very strong and very narrow ideas, an unsympathetic manner and a dogmatic way of laying down the law and refusing to depart from the letter thereof which is calculated to arouse suspicion . . . On the other hand, I am perfectly convinced that he is honest and that the accusation that the colony is a swindle is completely false.[9]

Honest or not, Lane was determined to follow a policy of 'isolation from a society whose seductions would draw our thoughts when they are needed to organise and build.'[10] Findlay should have concluded that Lane's problem was not deception, but delusion, since a man who so wholly believes in his own manifesto does not have the perspective to be playing a game of deceit. Lane was so convinced of his divine right to rule that he even appears to have taken it upon himself to arrange suitable marriages, something that had unhappy consequences for his old friend, Mary Gilmore, née Patterson, the budding Australian poet, who was his early supporter when he was still in Brisbane. Before her arrival, however, the community had split because of Lane's puritanical rules regarding alcohol (he was a lifelong tee-totaller) and native women. In 1894, Lane departed with forty-five adults and twelve children in tow to form another community further up the river. He named it Colonia Cosme. After the group had put in an arduous six months of clearing, planting and building, Mary Gilmore arrived in Cosme, committed to the cause and full of the enthusiasm she had regularly invested in the articles she wrote for the New Australia Association's journal, which she edited in Sydney. She could never have imagined how acutely embarrassing her arrival would be, when the man who Lane had 'organised' for her to marry not only showed no interest in her but was visibly contemptuous. He had been in love with someone else—the community's nurse, Clara Brown—but she had left him after Lane announced the 'betrothal' to Mary. It was typical of Lane's personal meddling that it resulted in broken hearts all around.[11]

---

9. Quoted in Anne Whitehead, *Paradise Mislaid* (Brisbane: University of Queensland Press, 1997), 208. For details of Lane's racist views see also JB Henderson 'William Lane, the Prophet of Socialism: The Tragic Paraguayan Experiment' read at a meeting of the Royal Historical Society, Queensland 27 June 1968: 530–31.

10. Whitehead, *Paradise Mislaid*, 200.

11. Anne Whitehead, *Blue Stocking in Patagonia* (London: Profile Books, 2003), 25.

Apart from this personal fiasco for Mary, which resulted in her marrying another man who was less than half her equal, life was hard and the food monotonous—even the robust Mary was shocked to discover that it consisted only of beans and the occasional monkey.[12] Illness was a constant problem, and it even got the better of William Lane and his wife, both of whom suffered from several health ailments. Because conditions were so difficult, people came and went, and eventually Lane himself had had enough of it. He left for London in 1899, and later took up a succession of newspaper jobs in New Zealand, where he would continue to promote communism. Mary Gilmore and her husband and child would depart soon after, and Cosme struggled on, but it had lost its ideological focus and racial purity. The descendants of Lane's original community, who were visited by ABC TV's *Foreign Correspondent* (19 September 2006), still remember their Australian origins, although their dark skin—a result of intermarriage with the indigenous population—would have disappointed Lane.

Curiously enough, William Lane was not the first to attempt a racist utopia in Paraguay. Seven years before Lane arrived, Elisabeth Nietzsche, the sister of the German philosopher Friedrich Nietzsche, travelled to Paraguay with her husband Bernhard Forster to establish a community that would enable 'the purification of the human race and the rebirth of a nation'. Forster, a notorious anti-Semite who had been fired from his teaching position in Berlin because of his racist views, was described in *The Times* of London:

> He is a man like too many of his countrymen, of one idea, and that idea is Germany for the Germans and not for the Jews. Finding that idea unrealistic in his native country, he, with a few devoted men like himself, has sailed to a far country, there to found a new Deutschland, where synagogues shall be forbidden, and Bourses [stock markets] unknown.[13]

But the Forsters' paradise was not viable, and soon fell into ruin. Bernhard Forster's grandiose delusions ended in a cocktail of strychnine and morphine, which he could face more readily than the humiliation of his failed utopia. It could not have happened to

---

12. Whitehead, *Blue Stocking in Patagonia,* 23.
13. Whitehead, *Paradise Mislaid,* 415.

a more deserving person. Elisabeth returned to her native Germany, where she nursed her dying brother Friedrich, who was suffering from mental illness. She edited his papers to reflect her own anti-Semitic views, with which he was unsympathetic, and forever tainted the German philosopher's reputation. Elisabeth became a darling of the National Socialists and a friend of Hitler; when she died in 1935, the SA, the SS and the Hitler Youth were all represented at her funeral.[14] The unsuccessful utopian experiment in Paraguay was but a negligible foretaste of the dystopia that would soon engulf Europe.

A steady stream of fantastic dreamers was drawn to the great unspoiled forests of South America in the hope of finding a direct path to Eden, a second chance at life before the Fall. The attraction was not always the moral and racial regeneration of the human race, however. There was also a decidedly material attraction to a continent where young men flocked to seek their fortunes and scoundrels lay in wait to dupe them with promises of untold bounty, cheap land and fail-safe investment opportunities. In the comic novel *The Adventures of Martin Chuzzlewit*, Charles Dickens satirised both the moral and the financial presumptions that usually accompanied these New World dreams and speculations. Martin, a young man from London who goes to America to seek his fortune, is soon tricked by the General, a member of the aptly named Eden Land Corporation, into forfeiting all his money for phony real estate:

> 'Well! you come from an old country; from a country, sir, that has piled up golden calves as high as Babel, and worshipped 'em for ages. We are a new country, sir; man is in a more primeval state here, sir; we have not the excuse of having lapsed in the slow course of time into degenerate practices; we have no false gods; man, sir, here, is man in all his dignity. We fought for that or nothing. Here am I, sir,' said the General, setting up his umbrella to represent himself, and a villanous-looking umbrella it was; a very bad counter to stand for the sterling coin of his benevolence, 'here am I with grey hairs, sir, and a moral *sense*. Would I, with my principles, invest capital in this speculation if I didn't think it full of hopes and chances for my brother man?'[15]

---

14. Whitehead, *Paradise Mislaid*, 419.
15. Charles Dickens, *The Adventures of Martin Chuzzlewit* (London: Thomas Nelson and Sons, 1900).

Rich in biblical imagery, the General's hard sell contrasts the moral turpitude of old Europe with the vision splendid of the New World, 'man in all his dignity' fully moral and interested in building up his brother man. If Thomas More had written *Utopia* three hundred years later he might have chosen this wryly comic style, for by the nineteenth century the grand hopes and visions of the Age of Exploration had surely given way to the unscrupulous schemes of an Age of Swindlers. In fact, William Lane was suspected of as much, although he appears to have been a genuine dreamer who did not profit by his utopian experiment; quite the reverse, he spent much of his life trying to pay back what the settlers had lost. Lane was perhaps a rare exception. In Dickens's time, however, there was a certain cause célèbre, which had begun in the 1820s and was still causing mischief well into the 1830s—and might just have inspired the Eden episode of Martin Chuzzlewit's misfortune.

At a time when America was opening up the western frontier and the California gold rush made ordinary men into tycoons overnight, when railroads were being laid and the American Indians were being steadily subdued, the New World positively gleamed with possibilities for the restless inhabitants of the Old World. Further south, however, the man known as the George Washington of South America, El Liberator Simon Bolivar, was fighting a war of independence from the Spanish, while the British were also fighting the Spanish over possession of the Caribbean islands. Back in England, the situation was dimly understood, being, in all frankness, a shambles, very poorly organised, with British soldiers arriving virtually without training, and unsuited to the terrain and climate. It was fertile ground for short-lived victories and tall tales, and one of them is investigated in David Sinclair's book *The Land That Never Was*.[16]

Among those who lived to tell the tale of his military adventures was one Gregor MacGregor. As was the custom of the day, he had paid for his military commission, and was neither a natural nor a gifted soldier, as the record of his military exploits makes painfully clear. However, MacGregor's personality inclined towards the grandiose, not unlike his sometime friend and ally, Simon Bolivar. Yet, unlike Bolivar, MacGregor was a soldier of fortune, on the lookout for land

16. David Sinclair, *The Land That Never Was: Sir Gregor MacGregor and the Most Audacious Fraud in History* (Cambridge, MA: Da Capo Books, 2003).

he could sell to investors at a pure profit; the war of independence raging around him was merely a means towards this end. After a largely undistinguished career, in which on more than one occasion he made a cowardly escape, abandoning his men to the enemy, MacGregor returned from Latin America to his native Scotland in 1820 bearing the title of Cazique (Prince) of Poyais, an independent country on the Bay of Honduras, which was reputedly free from the tumult that had engulfed the surrounding territories.

Back home, MacGregor was on a mission to entice investors and settlers to a country that was not only rich in natural resources but also had an established infrastructure, a beautiful main city, and houses waiting to be occupied. What is more, the native population was Anglophile and eager to become even more so. Armed with maps and freshly printed Poyais money, MacGregor's prospects were

Gregor MacGregor, a soldier of fortune, scammed his fellow Scotsmen into purchasing land in South America to join a thriving new colony based on the utopian dream. But when the colonists arrived in 1823 on the Mosquito Coast, they were in for a shock. Courtesy National Galleries of Scotland.

good in the commercial centres of Edinburgh and London, where he quickly found his willing investors. After all, South America had already captured the imagination of the British with the publication in the *Quarterly Review* of a series of articles on the jungle expeditions of the German natural scientist Baron Alexander von Humboldt. MacGregor's more commercially seductive invitation, along with his message that the natives were friendly, had come along just in time to exploit the growing public awareness of this far-off land.

MacGregor's advertising campaign garnered enough financial backing for him to charter two ships, the *Honduras Packet*, which sailed from London in 1822 with seventy passengers, and the *Kennersley Castle*, which sailed from Leith Harbour in Scotland in 1823 with about two hundred passengers. They were neither the poor nor the desperate, but a wide assortment of men and women, drawn from all classes and professions of society, each having been promised a role to play in a new country that was eager to accommodate them. Lawyers, doctors, cooks and a banker ensured that life in Poyais would be orderly, healthy and economically viable, with none of the social ills that were currently plaguing London and were grist to Charles Dickens's literary mill. The Poyais money they carried with them would be put to good use, and the maps portrayed a spacious settlement, with a lovely city named St Joseph that already boasted palatial buildings in the Spanish style, including a theatre, but which was now poised to be built up with an English infrastructure. With its temperate climate and hard-working natives who were used to low pay, St Joseph would soon become an important financial and cultural capital of Latin America.

When the *Kennersley Castle* anchored off that stretch of land called the Mosquito Coast (but which the passengers' guidebooks declared was really called Mosquitia) in the spring of 1823, they were perplexed to find that the city was nowhere in view. The map must have been a little inaccurate. A small boat was dispatched to check the depth of the harbour, but the crew was surprised to discover not a single ship nor sign of life. Eventually, on 22 March, the restless passengers were taken ashore. What awaited them was nothing but swampy, sandy land, with dense undergrowth and no passage visible through it. If there was a city it was not here. Desperation turned to despair, until some people were spotted some distance away, across a body of water that separated them from the mouth of the Black River, which had been clearly drawn on their maps. They soon realised that the others, who looked European, were the settlers who had arrived earlier on the *Honduras Packet*. They undoubtedly had come from St Joseph to greet them.

Upon meeting the welcoming party, the new arrivals were in for a further shock. To date, not a single city or town had been discovered. As for the native inhabitants, they had never heard of St Joseph or the Cazique of Poyais, nor did they have any use for the Poyais dollars which the new settlers thrust at them in the hope of buying provisions. Whatever life they would enjoy in this remote place, it would be at the most basic level of subsistence and at the expense of backbreaking work clearing, planting and hunting. The utopian dream had turned into a waking nightmare of squalor, disease and conflict. One settler, who had left a prosperous business behind as well as a family with whom he hoped to reunite in their new home, committed suicide in despair. Many more would succumb to malaria and other illnesses.

Fortunately, the settlers were accidentally discovered by the captain of the *Mexican Eagle*, from the English colony of Belize. When they were rescued and taken to the hospitals of Belize, it was too late for most of them, and about 180 died of deprivation and disease. Amazingly, back in London, MacGregor was still selling passage to his mythical Poyais, but fortunately a warning from the Superintendent for Belize, Edward Codd, resulted in the calling back of five more ships. About fifty of the original settlers returned to London from Belize in October 1823 aboard the *Ocean*, and their ordeal made exciting newspaper copy.[17] The perpetrator of this huge scam fled to France, but was later imprisoned in a Paris jail.

Although there are great differences in the story of Rajneeshpuram and Poyais, with the latter having never really got off the ground, there is a remarkable similarity in one sense. When the man who had been appointed 'governor' of Poyais, James Hastie, came to write his version of what he had believed to be a great utopian experiment in his book *Narrative of a Voyage in the Ship* Kennersley Castle *from Leith Roads to Poyais*, he did not blame Gregor MacGregor for the fiasco, but his advisers and publicists. It was they, he averred, who had spread false information, whereas MacGregor (he referred to him as

---

17. Sinclair notes that 'fewer than 50 ever saw Britain again' which may indicate some died on the return journey. *The Land That Never Was*, 240.

Sir Gregor) was a thoroughly noble man who would have certainly seen the scheme through to success if only he had accompanied them to Poyais. Hastie and some other survivors of the ordeal even signed a statement to that effect. It seems not to have occurred to them that MacGregor's avoidance of such a responsibility and his attempt to escape justice when his scheme was revealed as a baseless fraud were hardly the actions befitting a man of noble character.

However, the belief that MacGregor was not a liar and a cheat served him well, and after a brief stay in jail he went on to sell shares in equally dubious land prospects. He was, after all, the living symbol of a chance at utopia in a pure and paradisical place, belief in which was extremely attractive at a time when the industrialisation of England crowded the cities with the rural poor and polluted the air with grime and soot. Anyone offering a way out surely had his heart in the right place! This was precisely the same kind of argument mounted by many followers of Bhagwan Sri Rajneesh when Rajneeshpuram collapsed. All the blame was heaped on his advisers, while the guru, they claimed, was innocent of all charges. Despite the enormous fallout in the press at the time, which included graphic accounts of his erratic and controlling behaviour, the newly named Osho returned to India where his devoted following was not the least put off by his grandiose lifestyle, and his by-now debilitating addiction to large daily doses of nitrous oxide and Valium.

Innocence is a major ingredient of utopia, although usually it is the quality most often apparent in would-be followers. It is rarely associated with the leaders who engineer their communities and willfuly control their members' personal lives in every detail.

After all, as Bhagwan Sri Rajneesh made clear in the quote that opened this chapter, he knew his business very well—as his considerable wealth testified. Yet despite utopia's tendency to fail to live up to its promises, it nonetheless continues to have all the allure of buried treasure, there to be had if one could but find the map. If only its hapless seekers realised that such an impossible dream is signaled in the very name utopia, 'no place'. This is as it should be, since utopias only serve themselves, devoid of the spirit of universalism that enables one to see beyond particular doctrines and forms of social engineering to the value of other communities and cultures. One can only speculate how much poorer we would be if utopian experiments had snared more than the minorities they usually attract. Fortunately, the great Argentinean writer Jorge Louis Borges[18] was saved from putting his plans for an anarchist, free-love communitarian utopia into action in Paraguay when he fell head over heels in love with Leonor Acevedo. He decided in favour of his sweetheart instead, and married her. The world of letters is forever grateful.

---

18. Edwin Williamson, *Borges, A Life* (London: Penguin, 2005), 73.

Index